BY THE EDITORS OF CONSUMER GUIDE®
MEDICAL BOOK OF REMEDIES

50 Ways to Lower Your Fat & Cholesterol

IN ASSOCIATION WITH
THE TUFTS UNIVERSITY
SCHOOL OF NUTRITION

ROBERTA LARSON DUYFF, M.S., R.D., C.H.E.
CONSULTANT: JEANNE P. GOLDBERG, PH.D.

PUBLICATIONS INTERNATIONAL, LTD.

All rights reserved under International and Pan American copyright conventions. Copyright © 1996 Publications International, Ltd. This publication may not be reproduced or quoted in whole or in part by mimeograph or any other printed or electronic means, or for presentation on radio, television, videotape, or film without written permission from Louis Weber, C.E.O. of Publications International, Ltd., 7373 North Cicero Avenue, Lincolnwood, Illinois 60646. Permission is never granted for commercial purposes. Printed in U.S.A.

Neither the editors of Consumer Guide® and Publications International, Ltd., nor the author, consultant, or publisher take responsibility for any possible consequences from any treatment, procedure, exercise, dietary modification, action, or application of medication or preparation by any person reading or following the information in this book. This publication does not attempt to replace your physician or other health-care provider. Before undertaking any course of treatment, the publisher, author, and consultant advise you to check with your doctor or other health-care provider.

Contributors:

Roberta Larson Duyff, M.S., R.D., C.H.E., is a Registered Dietitian, Certified Home Economist, and writer whose work and articles have appeared in *Cooking Light, Better Homes and Gardens, Ladies' Home Journal,* and *USA Today.* She works actively with The American Dietetic Association and has served as a Board member and Foundation President of the Society for Nutrition Education and as Chairperson of the Home Economists in Business of the American Home Economics Association.

Jeanne P. Goldberg, Ph.D., is an Associate Professor at the Tufts University School of Nutrition and serves on the Editorial Advisory Board of the *Tufts' Diet and Nutrition Letter.* She was a research nutritionist with the Harvard School of Public Health and is on the Advisory Committee to the Food and Drug Administration Center on Food Safety and Applied Nutrition. She is a member of the American Public Health Association, the Society for Nutrition Education, and The American Dietetic Association.

The Tufts University School of Nutrition is uniquely positioned as the only School in the United States whose teaching and research programs fully cover all of nutrition science and policy. Its strength is its faculty and resources, representing the wide ranging areas of dietetics, food safety, and applied food policy, agriculture and the environment, clinical and preventive nutrition, and state of the art biological research using human subjects. These capabilities are employed to solve the nutritional problems of populations in the United States and abroad.

Cover photo: FPG International

CONTENTS

Introduction 7

SETTING YOUR GOALS

1	Take This Message to Heart	9
2	Learn Fat Lingo	13
3	Have Your Blood Cholesterol Checked Regularly	17
4	Rate Your Plate	19
5	Keep the Dietary Guidelines in Mind	21
6	Pyramid Your Food Choices	26
7	Figure Your Fat Budget	30
8	Make Fat Trade-Offs	34

AT THE SUPERMARKET

9	Know Label Lingo	36
10	Catch Ingredient Clues	41
11	Buy Lean at the Meat Counter	43
12	Plan Fishy Meals	46
13	Remember the Five-A-Day Advantage	48
14	Go for Grains	51
15	Make Wise Dairy Decisions	53
16	Try Modified Versions of Traditional Favorites	56

IN YOUR KITCHEN

17	Trim the Fat	58
18	Go Skinless	59
19	De-Fat Your Cooking Style	60
20	Use a Light Hand with Sauces and Such	63

21	De-Grease Soups and Gravies	66
22	Alter Recipes to Stretch High-Fat Ingredients	67
23	Lose Fat, Not Flavor	68
24	Make Easy Substitutions	71
25	Watch Portion Size	72
26	Be "Eggs-tra" Smart	74

SNACKS AND MEALS

27	Control Your "Fat Tooth"	76
28	Graze on Smart Snacks	77
29	Boost Your Fiber Factor	78
30	Go Vegetarian Every Now and Then	80

EATING OUT

31	Control Fat in the Fast-Food Lane	83
32	Order Wisely	86
33	Look for Menu Clues	88
34	Give Your Taste Buds a Passport	89

AN ACTIVE LIFESTYLE

35	Get Moving	94
36	Test Your Exercise Quotient	99
37	Choose an Activity Just Right for You	100
38	Plan an Exercise Program for F.I.T.	101
39	Walk Your Way to Health	103
40	Fit in Unscheduled Exercise	105
41	Put Action in Your Workday	107
42	Give Your Body a Regular Workout	109

| 43 | Make Exercise a Family Affair | 112 |
| 44 | Enjoy Active Vacations | 113 |

OTHER FACTORS

45	Keep Your Healthy Weight	114
46	Be Extra Careful After Menopause	118
47	Drink Wine in Moderation Only	120
48	Smokers—Kick the Habit	121

LIFETIME SUCCESS

| 49 | Ask the Experts | 122 |
| 50 | Savor the Results | 124 |

Index 126

INTRODUCTION

Many factors can contribute to the development of the nation's number one killer—heart disease. Some of these factors, such as heredity and age, are beyond your control, but others are not. Research shows that certain lifestyle habits have a major impact on cardiovascular health. Reducing fat and cholesterol is one step that can make a difference. Diet, exercise, and personal habits are all part of the fat and cholesterol equation and all areas of your life over which you have direct control.

Fat and cholesterol reduction means several things. Lowering fat means both reducing dietary fat and maintaining a healthy weight by keeping an eye on body fat. Lowering cholesterol means both cutting dietary cholesterol and controlling the level of cholesterol in your bloodstream. This book can help you attain all these goals.

With all the health news about cutting fat, consumers sometimes forget that fat is an essential nutrient. It's a concentrated source of energy (or calories) and carries the vitamins A, D, E, and K in food and in the

INTRODUCTION

bloodstream. Fat also offers satiety, or a feeling of fullness after eating.

Today's concern about fat is not its value as a nutrient, though; it's the excessive amount most people consume. About 35 percent of the calories in a typical American diet come from fat. Health professionals now tell us that no more than 30 percent of total calories should come from fat. For most of us this means simply moderating our eating habits and choosing mostly lower-fat foods.

It is important to remember not to cut out any category of food in an effort to avoid fat; you may end up missing out on other nutrients. Also, for infants and children, restricting fat isn't advised. Fat provides essential fatty acids that they need for normal cell growth and development.

This book can help you navigate a low-fat course. From wise food choices and lean cooking techniques to getting the whole family involved in physical exercise, you will learn the way to a healthy lifestyle. It's really not as hard as you think.

SETTING YOUR GOALS

1

TAKE THIS MESSAGE TO HEART.

The first step toward lowering your fat and cholesterol is knowing why it's important and what's at stake if you don't. Learning about fat and cholesterol will help you understand why health professionals are concerned and why you should be, too.

Diets high in fat, especially the kind of fat called *saturated fat,* are linked to high levels of cholesterol in the blood—a condition that is a serious risk factor for heart disease. As blood cholesterol levels go up, the risk of heart disease goes up, too. Obesity and some cancers are other health problems linked to high-fat diets.

Foods have a mixture of several types of fat: Fats in beef, pork, dairy foods, and egg yolks are mainly saturated fats; the fat in seafood is mainly polyunsaturated; and the fat in poultry, nuts, and olives is mostly monounsaturated (see definitions in tip 2). Among fats and oils, lard, butter, and tropical oils (palm, palm kernel, and coconut oils) are mostly saturated. Corn, cottonseed, sunflower, safflower, and soybean oils are

SETTING YOUR GOALS

mostly polyunsaturated fat; and olive, canola, and peanut oils are mostly monounsaturated fat. These distinctions will be important later.

Cholesterol is a fat-like substance found in food. It is part of every cell in the body and many hormones. In varying amounts, cholesterol comes from animal sources of food but not from fruits, vegetables, or grains. The body makes it, too, which is why it is not an essential part of the diet. We make all the cholesterol we need. If our bodies make too much, our risk of heart disease goes up.

Blood cholesterol isn't just one substance either. In the blood, cholesterol is linked to lipoproteins, either high-density lipoproteins (HDLs) or low-density lipoproteins (LDLs). HDLs help transport cholesterol away from body tissues so they can be removed as waste. LDLs help carry cholesterol to body tissues, including the walls of the arteries, and here is where the true danger lies. Cholesterol deposits on the walls of the arteries that supply the heart cause the blood vessels to become narrow and even blocked, thus cutting off the vital flow of oxygen to the heart muscle and causing a heart attack. The blood vessels to the brain are also susceptible to this narrowing. Blockages in these vessels cause parts of the brain to be cut off from their oxygen supply, resulting in strokes. Because LDL cholesterol is the kind that precipitates the buildup that causes blockages, it is often called bad cholesterol; because HDL cholesterol is headed out of the system, it is often called good cholesterol.

Heredity, age, body weight, exercise level, and smoking all make a difference in the amount of cholesterol in our blood. To some degree, cholesterol in food does, too. That's why health experts recommend limiting dietary cholesterol intake to 300 milligrams a day. But the biggest dietary influence on blood cholesterol levels is not dietary cholesterol—it's dietary fat.

A diet high in saturated fat elevates blood cholesterol levels and appears to lower blood HDL cholesterol levels. Polyunsaturated and monounsaturated fats work differently. Polyunsaturated fat tends to lower both HDL and LDL cholesterol levels. Monounsaturated fat doesn't appear to affect HDL cholesterol levels, but does help lower LDL cholesterol levels.

Even though mono- and polyunsaturated fats are better than saturated fat, health experts advise: Eat less total fat. Every type of fat has the same calorie bottom line—nine calories per gram—meaning that too much of any of them can lead to weight gain.

Another fat to watch out for is *trans* fatty acids, which are made by changing the structure of unsaturated fat through hydrogenation. Hydrogenation—the process of adding hydrogen to liquid oils—makes them more solid, stable, and saturated. This change causes the previously unsaturated fat to act more like saturated fat, helping to raise total blood cholesterol levels. Some of the fat in margarine is the *trans* fatty acid type.

For infants and children, restricting fat isn't advised. Fat provides essential fatty acids (linoleic and linolenic) that children need for normal cell growth and development. By about age two, when they're ready for family foods, caregivers should urge moderation. Children's diets should not be too restricted, though; they may not get enough calories and nutrients for growth and energy.

SETTING YOUR GOALS

2

LEARN FAT LINGO.

Learning the language that doctors and dietitians use is a big step toward understanding the roles of fat and cholesterol in food and health. You don't have to memorize these terms, but as you read the rest of the book, you may refer to this glossary as needed.

Cholesterol: substance that resembles fat, but in the body has a chemical structure and function different from fat. We talk about cholesterol in several ways:

Blood cholesterol (or **serum cholesterol**) is the cholesterol that circulates in the bloodstream.

Dietary cholesterol is the cholesterol found in food. Cholesterol comes only from animal sources of food; egg yolks, organ meats, meat, and poultry are the main sources; some shellfish and whole-milk dairy foods also contain significant amounts. Because the body makes all the cholesterol it needs, it is not an essential nutrient.

High-density lipoprotein (HDL) cholesterol is blood cholesterol carried by lipoproteins that have more protein than

SETTING YOUR GOALS

fat. This type of cholesterol is often referred to as "good" cholesterol.

Low-density lipoprotein (LDL) cholesterol is blood cholesterol carried by lipoproteins that have more fat than protein. This type of cholesterol is often referred to as "bad" cholesterol.

Fat: group of compounds made of one or more fatty acids and glycerol. When fats from food are digested in the body, they first break down into their component parts. Then, the body changes them into lipoproteins, so they can be carried in the bloodstream. Like cholesterol, fat is referred to in several ways:

Adipose tissue is the scientific name for body fat. Our body stores energy mainly in the form of fat. Body fat also protects vital organs and insulates the body from heat and cold.

Dietary fat is the fat found in food. Dietary fat is a nutrient that provides energy, carries vitamins A, D, E, and K, adds to the satiety of food, and provides essential fatty acids. Essential fatty acids cannot be produced by the body and must be obtained through dietary fat; they are especially important for growing children.

Triglycerides are the major storage form of fat found in the body. Most of the fat found in food is in the form of triglycerides.

Fatty acids: building blocks of fat. Fatty acid molecules are chains of carbon, hydrogen, and oxygen. Fatty acids differ from one another in two ways: the length of the carbon chain and the degree of saturation. Most fats are made of mixtures of the three types of fatty acids (monounsaturated, polyunsaturated, and saturated):

Monounsaturated fatty acids are missing one pair of hydrogen atoms on the chain (hence the "mono"). Canola, olive, and peanut oils contain mostly monounsaturated fatty acids.

Polyunsaturated fatty acids are missing two or more pairs of hydrogen on the chain. Plant oils—corn, cottonseed, sesame, safflower, and soybean oils—and seafood are good sources. These fats are liquid at room temperature. Most vegetable oils are polyunsaturated fats.

Saturated fatty acids have hydrogen linked to all the carbons in the chain. Saturated fatty acids are found mainly in animal fats, such as beef fat and butter fat; however, some vegetable fats, such as palm, palm kernel, and coconut oils, are highly saturated, too.

Omega-3 and omega-6 fatty acids are two kinds of polyunsaturated fatty acids. Their chemical structure differs slightly. Omega-3 fatty acids are found mainly in seafood, especially cold-water fish, and they may help protect against heart disease and stroke.

Trans **fatty acids** are polyunsaturated fatty acids that have undergone hydrogenation to make the fat solid or semi-solid. Some studies suggest that *trans* fatty acids may raise blood cholesterol levels.

Hydrogenation: processing that adds hydrogen to unsaturated fatty acids, making them more saturated and more firm.

Lipoproteins: body chemicals made of fat and protein. Some lipoproteins carry cholesterol in the bloodstream.

SETTING YOUR GOALS

3

HAVE YOUR BLOOD CHOLESTEROL CHECKED REGULARLY.

What's your blood cholesterol level? If you don't know, it's a good idea that you find out. The average American adult has a blood cholesterol level that's between 205 and 215 milligrams per deciliter (mg/dL). That number indicates the total concentration of cholesterol in the blood. Health professionals advise us to keep our blood cholesterol levels under 200 mg/dL to reduce the risk of heart disease. Although there are no hard cutoffs, and different individual factors make absolute numbers difficult to pin down, for most people, 200 to 239 mg/dL is considered borderline high; 240 mg/dL and over is high.

After age 20, you're smart to have your blood cholesterol checked at least once every five years—more often if you're older or at risk for cardiovascular disease. To get a more complete picture, your doctor may measure HDL and LDL cholesterol levels separately and check your triglyceride level, too, as part of a physical exam.

Cholesterol screenings give approximate blood cholesterol levels. Using a drop of blood taken from your finger, a portable machine analyzes the sample within minutes. However, these tests—often given in shopping malls, supermarkets, and health fairs—may not be as accurate as tests done with better techniques and equipment and with more highly trained health professionals in your doctor's office or health maintenance organization. So if you're at risk—or if your number comes up high or border-line high—have another blood sample checked by your physician.

Why have triglycerides checked? Some researchers believe that blood triglyceride levels may provide additional information in determining risk, and elevated triglyceride levels often go hand-in-hand with low levels of HDL cholesterol. High blood triglyceride levels may also suggest diabetes and kidney disease.

Desirable Blood Levels

Total blood cholesterol less than 200 mg/dL

Blood LDL cholesterol less than 130 mg/dL

Blood HDL cholesterol more than 35 mg/dL

Blood triglycerides less than 250 mg/dL

SETTING YOUR GOALS

4

RATE YOUR PLATE.

How would you judge your diet? Do you "lean toward health" by controlling the fat in your food choices? Always or sometimes? Do you balance out your diet with wise choices? Consider the following questions and circle the answer (**A** for always, **S** for sometimes, **N** for never) that best describes your eating habits. Answer them truthfully; you only stand to help yourself. Do you. . .

Select lean cuts of meat?	A	S	N
Trim visible fat from meat?	A	S	N
Remove the skin before eating poultry?	A	S	N
Enjoy fish and poultry as your main dish several times a week?	A	S	N
Keep your portions of meat, poultry, and fish to 3 to 5 ounces?	A	S	N
Broil, roast, or braise meat, fish, or poultry rather than fry it?	A	S	N
Skim fat from soup, stew, and broth?	A	S	N
Eat five servings of fruits and vegetables a day?	A	S	N
Use high-fat dressings sparingly?	A	S	N
Usually enjoy fruit for dessert, leaving pastries or cakes for special occasions?	A	S	N

SETTING YOUR GOALS

Limit egg yolks to four per week, including those in baked foods? A S N

Eat at least six servings of grains per day? A S N

Eat low-fat and fat-free dairy foods? A S N

Make a point of eating fiber-rich foods? A S N

Read food labels to compare the fat content of food? A S N

Eat mostly low-fat foods, leaving room to enjoy some higher-fat foods, too? A S N

Count the circles in each column. What do your answers suggest about your diet? If most of your answers fall in the "never" category, you have a way to go. If "sometimes" was your most frequent answer, you still have room for improvement. Now that you know where you stand, you can make specific changes to meet your individual needs.

5

KEEP THE DIETARY GUIDELINES IN MIND.

Healthy eating is much more than fat and cholesterol control. It's a way of eating that blends a wide variety of healthful foods into a balanced, moderate diet—with great taste and satisfaction along the way!

The Dietary Guidelines for Americans paint the picture of good food choices for healthy Americans aged two and over. These seven recommendations from the U.S. Department of Agriculture and the U.S. Department of Health and Human Services represent the latest advice of nutrition experts. These guidelines are designed to help you not only lower the amount of fat and cholesterol in your diet, but also insure that you get all the nutrients you need.

- **Eat a variety of foods.**

To keep fit, your body needs all kinds of nutrients, and since no one food supplies all nutrients in an adequate amount, you need variety.

Within that variety, you'll find all the foods you enjoy. The good news is that there are no good or bad foods, just good or bad diets. Any food can fit as long as the total diet over a day or more meets the Dietary Guidelines. (A varied diet follows the Food Guide Pyramid, shown in tip 6.)

- **Maintain a healthy weight.**

Your best weight—not too fat or too thin—is one of your best assets for fitness now and in the long run. Too much body fat is linked to many health problems, including heart disease, stroke, cancer, and diabetes.

What's your best weight? Refer to tip 45, "Keep Your Healthy Weight," to see how you can determine your own healthy weight. Then, if you need to lose or gain, set reasonable goals, and reach them with a healthful diet and plenty of exercise.

- **Choose a diet low in fat, saturated fat, and cholesterol.**

This guideline is the main focus of our endeavor to lower fat and cholesterol. Americans are advised to limit fats, especially saturated fats, and cholesterol in their food choices and so decrease their risk of heart disease, some cancers, and obesity. (Tip 7, "Figure Your Fat Budget," shows you how to set your own fat, saturated fat, and cholesterol targets.)

SETTING YOUR GOALS

- **Choose a diet with plenty of vegetables, fruits, and grain products.**

Vegetables, fruits, and grain products are great sources of complex carbohydrates, fiber, and vitamins and minerals. Fruits and vegetables supply varying amounts of vitamins A and C, for example—both important antioxidant vitamins. Grain products such as bread, pasta, rice, and barley supply thiamin, niacin, and iron. They all tend to be low in fat, and remember, foods from plant sources have no cholesterol. (Tips 13 and 14, "Remember the Five-A-Day Advantage" and "Go for Grains," offer many ideas for adding fruits, vegetables, and grain products to your diet.)

- **Use sugars only in moderation.**

Why moderate sugars? They're found naturally in many foods. But sugars are also an added ingredient, often found in foods that supply calories but few nutrients and in processed foods that also happen to be high in fat. Sugar is also a cause of tooth decay.

Sugar goes by many different names that you may not recognize unless you look for them: table sugar (sucrose), brown sugar, corn sweetener, fructose, dextrose, glucose, honey, lactose, molasses, maltose, and

SETTING YOUR GOALS

syrup. Find them on the ingredient lists of food labels and look for the total amount per serving so you can keep track of the total amount in your diet. (For more information on using food labels, see tip 9.)

- **Use salt and sodium only in moderation.**

Sodium is a nutrient. Yet most Americans consume more than they need. Some are sodium-sensitive; their blood pressure goes up as their sodium intake goes up. Cutting back on sodium and salt (sodium chloride) is one way to control blood pressure. (High blood pressure is a risk factor for cardiovascular disease.)

Most of the sodium in our diets comes from salt, used either in processing or in cooking, but many other ingredients have sodium: baking powder, baking soda, monosodium glutamate (MSG), seasoned salt, sodium benzoate, sodium citrate, sodium nitrite, sodium phosphate, and soy sauce. Use food labels to compare the sodium content, then most of the time choose foods with less sodium. Limit the amount of salt you use at the table and in cooking; use herbs and other seasonings instead. Cut back on salty snacks, such as

SETTING YOUR GOALS

chips, pretzels, and salty nuts. And be aware of foods that tend to be higher in sodium, such as processed meat, condiments, and prepared foods and cheeses.

- **If you drink alcoholic beverages, do so in moderation.**

Alcoholic beverages supply few, if any, nutrients, but they have plenty of calories. Although alcohol may have some health benefits, these are easily offset by its potential negative effects: Alcohol consumption is linked to accidents and health problems.

Some people shouldn't drink alcoholic beverages at all: children and teens, pregnant women and those who are trying to conceive, people who take certain medication (including some over-the-counter drugs), people who plan to drive or need skill and attention in what they do (for example, in the use of power tools), and those who cannot control their drinking.

Moderation means no more than one drink a day for women, and two, for men. (The discrepancy takes into account physiologic differences in the way men and women metabolize alcohol.) A drink is 12 ounces of regular beer, 5 ounces of wine, or 1½ ounces of distilled spirits (80 proof).

6

PYRAMID YOUR FOOD CHOICES.

Eating healthy has gotten a lot simpler with the Food Guide Pyramid. Its simple guidelines help you get the nutrients you need and the foods you like best. The practical approach of the Pyramid makes choosing a healthful diet easy. It offers a set of guidelines, not a rigid set of rules.

The Food Guide Pyramid is a way to implement suggestions found in the Dietary Guidelines for Americans you read about in tip 5. It shows you how to eat a varied diet

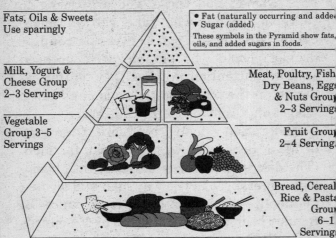

Fats, Oils & Sweets
Use sparingly

● Fat (naturally occurring and added)
▼ Sugar (added)

These symbols in the Pyramid show fats, oils, and added sugars in foods.

Milk, Yogurt & Cheese Group
2–3 Servings

Meat, Poultry, Fish, Dry Beans, Eggs & Nuts Group
2–3 Servings

Vegetable Group 3–5 Servings

Fruit Group
2–4 Servings

Bread, Cereal, Rice & Pasta Group
6–11 Servings

with all the essential nutrients, without consuming too many calories or too much fat, especially saturated fat, and how to get foods from different groups in the recommended proportion—more grains than vegetables, more vegetables than meats.

Each food group suggests a range of servings. The number of servings you need is an individual matter, based on your calorie needs, which depend on your age, sex, body size, and activity level. At the very least, eat the minimum number of servings for each food group. (See the chart on page 28 to figure your individual serving numbers.)

To capture the power of the Pyramid, you must also know what equals one serving (see the chart on page 29). Then, if your portion is larger, count it as more; if it's smaller, less. There are no serving sizes for the foods at the Pyramid tip. However, in small amounts, fats, sweets, and oils do add flavor and enjoyment to your meals.

How Many Servings for You?

	Many women, older adults	Children, teen girls, active women, most men	Teen boys, active men
Calorie Level*	about 1,600	about 2,200	about 2,800
Bread Group	6	9	11
Vegetable Group	3	4	5
Fruit Group	2	3	4
Milk Group	2–3†	2–3†	2–3†
Meat Group	2 for a total of 5 ounces	2 for a total of 6 ounces	3 for a total of 7 ounces

* You can reach these calorie levels by eating lower-fat, lean foods from the five food groups and by choosing fats, oils, and sweets in small amounts.

† Teenagers, young adults to age 24, and women who are pregnant or breast-feeding need three servings.

SETTING YOUR GOALS

How Much Is a Serving?

Food Group	Serving Size
Milk, Yogurt, and Cheese	1 cup of milk or yogurt 1½ ounces of natural cheese 2 ounces of processed cheese
Meat, Poultry, Fish, Legumes, Eggs, and Nuts	2–3 ounces of cooked lean meat, poultry, or fish (1 egg, 2 tablespoons of peanut butter, or ½ cup of cooked dry beans equals 1 ounce of meat)
Vegetables	1 cup of raw leafy vegetables ½ cup of cooked or chopped vegetables ¾ cup of vegetable juice
Fruit	1 medium apple, banana, or orange ½ cup of chopped, cooked, or canned fruit ¾ cup of fruit juice
Bread, Cereal, Rice, and Pasta	1 slice of bread 1 ounce of ready-to-eat cereal ½ cup of cooked rice, cereal, or pasta

SETTING YOUR GOALS

7

FIGURE YOUR FAT BUDGET.

Your body needs dietary fat, but far less than what you are probably eating. Just how much is enough—but not too much? If you're an average American, you consume about 34 percent of your calories from fat. Health experts advise most people to cut back on fat and control cholesterol by

- Reducing total dietary fat intake to 30 percent or less of total calories

- Reducing saturated fat intake to less than 10 percent of calories

- Reducing cholesterol intake to less than 300 milligrams per day.

The goal isn't to eliminate fat, or foods with fat. After all, fat is an important nutrient, and it tastes good. Instead, you need to control the amount of fat in your diet. When you want to control your spending, you figure a budget; when you want to control your fat intake, figure your fat budget.

Your actual fat budget—how many grams of fat you have slated for each day to get less than 30 percent of calories from fat—depends on your overall calorie needs. That's determined by your age, sex, body size, and activity level (see chart page 32). The goal is less than 30 percent of your calories from fat and only 10 percent of your calories from saturated fat. How much fat is that in a day? Here's how you figure it out:

Pretend, for example, that you need about 2,200 calories a day to maintain your current body weight. Here's how you would figure your target fat and saturated fat intake:

Total fat:

1. Thirty percent of 2,200 calories is 660 calories:
   $0.30 \times 2{,}200$ calories = 660 calories

2. Fat has 9 calories per gram. So the amount of fat that supplies 660 calories is about 75 grams:
   660 calories ÷ 9 calories/gram = 73.3 grams (or about 75 grams)

Saturated fat:

1. Ten percent of 2,200 calories is 220 calories:
   $0.10 \times 2{,}200$ calories = 220 calories

2. Fat has 9 calories per gram. So the amount of fat that supplies 220 calories is about 25 grams:

220 calories ÷ 9 calories/gram = 24.4 grams (or about 25 grams)

Now, figure the maximum fat, saturated fat, and cholesterol in your own daily diet.

Figuring Your Calorie Allotment

Select your activity level (low if you're relatively sedentary; moderate if you get some exercise; high if you exercise regularly), and find your level on the chart.

\multicolumn{4}{c}{Women}			
Age	Low Activity	Moderate Activity	High Activity
19–24	1,800	2,200	2,600
25–50	1,800	2,200	2,600
51+	1,700	2,000	2,400

\multicolumn{4}{c}{Men}			
Age	Low Activity	Moderate Activity	High Activity
19–24	2,300	3,000	3,700
25–50	2,300	3,000	3,800
51+	2,000	2,600	3,200

Reprinted with permission from Recommended Dietary Allowances, 10th edition. © 1989 by the National Academy of Sciences. Courtesy of the National Academy Press, Washington, D.C.

SETTING YOUR GOALS

Daily Calorie Level	1,600	2,200	2,800
Maximum Fat Intake (30% of calories)	53 g*	73 g	93 g
Maximum Saturated Fat Intake (10% of calories)	17 g	24 g	31 g
Maximum Cholesterol Intake	300 mg	300 mg	300 mg

* Rounding off these figures will make them easier to remember and won't significantly alter your budget. It would be almost impossible to calculate to the nearest gram exactly how much fat you consume; calculations to the nearest 5 grams are easier.

Based on recommendations of the National Academy of Sciences.

Now that you know your fat budget, the key is to eat within your means. Use the information on the labels to get an estimate of your fat intake. Are you within your budget? No? Well, read on to learn some strategies for living within your fat budget.

GOALS

8

MAKE FAT TRADE-OFFS.

Eating within your fat budget is like living within your household budget. You have just so much to spend. If, for example, a luxury vacation costs more, you must spend less on other things back home. With fat budgeting you can take a "luxury vacation" with your food choices—as long as you balance them with "economy" later on.

Fat trade-offs help you eat within your fat budget and even enjoy some higher-fat foods once in a while. You don't need to eat only low-fat foods if you plan your diet wisely. For example, you want a thick, creamy milk shake. Fries sound good to you, too. Choose what's more important—perhaps the shake. Enjoy a baked potato with salsa, which has no fat, and order the milk shake in your favorite flavor.

You can handle your budget in one of two ways. You can choose foods that spread your allotted grams of fat throughout the day, or you can eat lean most of the day, saving up for one dessert or snack you can't resist. The choice of how to spend your fat budget is yours. Perhaps most of the time, you spread

SETTING YOUR GOALS

your allotment throughout the day, but if you're planning to go out one evening, you switch gears and spend that day saving up.

Like your household budget, figure your fat budget over several days. If you overspend one day, trade off to cut back on fat the next day. Every day need not include your exact allotment of fat: Some days you go over; some days you stay under. The bottom line is 30 percent or less of calories from fat averaged over several days, not one meal or one day.

Try to be smart with trade-offs. Spending most of your fat budget on dairy and meat products rather than foods from the Pyramid tip, such as candy, is a wise choice; meat and dairy offer many more nutrients.

Control your diet with "cholesterol trade-offs," too. Unless you're on a special diet, your cholesterol budget should top out at 300 milligrams daily, averaged over several days. If you can't resist a double-egg omelet with at least 430 milligrams cholesterol, balance out that choice by avoiding eggs for a couple of days.

KNOW LABEL LINGO.

What's in the food package? Check the label! With its new format, the food label is our best resource for nutrient content. Food labels must appear on most packaged foods in the supermarket. This same labeling information is also provided voluntarily on many fresh foods—meat, poultry, fish, fruits, and vegetables.

Use Nutrition Content Claims

"Light," "cholesterol free," "rich in," "lean," and "healthy"—when you read a label, start with nutrition content claims. They appear on the front of many food containers. Although content claims don't give exact nutrient or calorie amounts, they're great for helping you quickly find foods with nutrition qualities you want.

In the past, nutrient content claims were often misleading because their meanings weren't consistent. Now terms are defined clearly. They may appear only on foods that meet strict guidelines. Use nutrition content claims as a quick reference. Then, for more specifics, check the Nutrition Facts panel.

AT THE SUPERMARKET

If the label says...	Then the food has...
Fat free	Less than 0.5 grams of fat
Saturated fat free	Less than 0.5 grams of saturated fat, and *trans* fatty acids make up 1 percent or less of the total fat
Low fat	3 grams or less of fat
Reduced or less fat	At least 25 percent less fat
Reduced or less saturated fat	At least 25 percent less saturated fat
Lean	Less than 10 grams of fat, 4 grams or less of saturated fat, and 95 milligrams of cholesterol or less
Extra lean	Less than 5 grams of fat, 2 grams of saturated fat, and 95 milligrams of cholesterol or less
Light (or lite)	⅓ fewer calories or 50 percent less fat
Cholesterol free	Less than 2 milligrams of cholesterol and 2 grams or less of saturated fat
Low cholesterol	20 milligrams or less of cholesterol per serving and 2 grams or less of saturated fat
Reduced cholesterol	At least 25 percent less cholesterol and 2 grams or less of saturated fat

AT THE SUPERMARKET

What's on the Nutrition Facts Panel?

The Nutrition Facts panel gives an in-depth view of each food. It offers great information for making food choices, comparing one food with another, and planning your meals and snacks.

Serving size. Serving sizes on the new label are realistic, given in amounts people often eat. Nutrients and calories in one serving are based on these amounts. If your serving is bigger, the nutrients and calories are more, too. (The serving size on a label may differ slightly from the recommended serving on the Food Guide Pyramid.) Check the serving size before you compare two similar foods. The amount may differ slightly.

Servings per container. This shows how many servings the container of food supplies.

Calorie information. This information helps you fit the food within your own calorie budget and tells how many calories come from fat. Remember, your whole diet counts, not the calories or percent of calories from fat in one food or meal.

List of nutrients. Rather than list all the nutrients your body needs, the Nu-

AT THE SUPERMARKET

trition Facts panel gives only those that relate to today's major health concerns. In the past, the list of vitamins and minerals was longer; now only vitamins A and C, calcium, and iron are required to appear on the food label.

Nutrition Facts Serving Size ½ cup (114g) Servings per Container 4	
Amount Per Serving	
Calories 90	**Calories from Fat** 30
	% Daily Value*
Total Fat 3g	5%
Saturated Fat 0g	0%
Cholesterol 0mg	0%
Sodium 300mg	13%
Total Carbohydrate 13g	4%
Dietary Fiber 3g	12%
Sugars 3g	
Protein 3g	
Vitamin A 80% •	Vitamin C 60%
Calcium 4% •	Iron 4%

*Percent Daily Values are based on a 2,000-calorie diet. Your daily values may be higher or lower depending on your calorie needs.

	Calories:	**2,000**	**2,500**
Total Fat	Less than	65g	80g
Sat Fat	Less than	20g	25g
Cholesterol	Less than	300mg	300mg
Sodium	Less than	2,400mg	2,400mg
Total Carbohydrate		300g	375g
Fiber		25g	30g
Calories per gram:			
Fat 9 •	Carbohydrate 4	•	Protein 4

AT THE SUPERMARKET

% Daily Values (%DVs). The %DVs show how one serving fits into a 2,000-calorie diet. They help you easily compare one food with another and determine if a food is high or low in specific nutrients.

For fat, saturated fat, cholesterol, and sodium, 100 percent is the maximum amount, but it's okay to have less. The %DVs can be your aid for budgeting your intake of total fat, saturated fat, cholesterol, and sodium over one or more days. Of course, you can always stick with counting the number of grams for your budget, but some people find the %DV easier.

Daily Values footnote. The footnote at the bottom of the label tells the Daily Values at two calorie levels: 2,000 and 2,500. Some, such as fat and cholesterol, are maximums. Others, such as carbohydrate, are minimums. Your own target may be higher or lower, depending on your calorie needs. This footnote is the same on every container.

Calorie conversion information. The calories per gram show that fat has more than twice the calories per gram than carbohydrate and protein.

AT THE SUPERMARKET

10

CATCH INGREDIENT CLUES.

You've read the Nutrition Facts panel, which tells how many grams of total fat and saturated fat a food has. Check the food label again. The ingredient list on the Nutrition Facts panel tells you exactly what's inside. The list is in order by weight, from most to least; so the ingredients present in the largest amounts appear first. Here are some specific things to look for.

Many foods, including prepared foods, crackers, cookies, and margarine, are made with some *partially hydrogenated vegetable oil*. That means that the oil is processed to be solid or semisolid, to prolong its shelf life, and perhaps to make it more spreadable. Hydrogenation creates *trans* fatty acids, which seem to have almost the same effect on blood cholesterol levels as saturated fat. If the first fat listed is hydrogenated vegetable shortening, then you know that a good deal of that product's fat is in the form of *trans* fatty acids.

The label gives the specific type of fat or oil used in the product. If it says *vegetable oil,* then the specific oil is given right after

AT THE SUPERMARKET

in parentheses, such as *soybean and/or cottonseed oil* or *coconut and/or palm kernel oil*. Three vegetable oils—coconut, palm, and palm kernel—are more saturated than other vegetable oils. These oils, sometimes called tropical oils, can be found in many processed baked goods. If you have a choice between several different products, choose the one that uses monounsaturated (canola, olive, peanut) or polyunsaturated (corn, cottonseed, sesame, safflower, soybean) oils instead of the saturated tropical oils or hydrogenated vegetable shortenings.

When you catch high-fat or high-cholesterol ingredients on the ingredient list, ask yourself, are these ingredients essential to the product? For example, refried beans are often made with lard. Vegetarian-style refried beans without the lard are virtually the same, but they are missing all of the cholesterol and saturated fat.

Train your eye to catch the ingredients that mean hidden fat or saturated fat. Be on the lookout for words like *hydrogenated,* and learn to recognize the tropical oils. Reading the ingredient list can steer you clear of the fat, saturated fat, and cholesterol that sometimes hide in food products.

AT THE SUPERMARKET

11

BUY LEAN AT THE MEAT COUNTER.

Both the breeder and the butcher are making today's cuts of meat and poultry leaner than ever. But you still need shopping savvy to buy lean at the meat counter.

- Get the "skinny" on meat. Three-ounce cooked and trimmed portions of the following lean cuts supply under 200 calories and contain about 8 grams of total fat, about 3 grams of saturated fat, and 80 milligrams of cholesterol or less:

  Beef: round tip, top round, eye of round, top loin, tenderloin, sirloin

  Pork: tenderloin, boneless top loin chop, cured boneless ham (regular and 95% lean), center loin chop

  Poultry: chicken breast (skinless), chicken leg (skinless), turkey light meat (skinless), turkey dark meat (skinless)

  Lamb: loin chop, leg

- Look for lean ground beef. Ground sirloin is leanest, followed by ground round, ground chuck, and finally, regular ground

AT THE SUPERMARKET

beef. Labeling offers other clues for ground meat. Any meat labeled "85 percent lean" or more is a good choice.

- Buy meat that's trimmed to one-quarter inch or less. There's no need to pay for fat you'll trim away when you get home.

- Choose meat with less marbling, or fat flecks, in the lean muscle. Grades of meat labeled "USDA Select" have less fat than "Choice" or "Prime." But they're equally high in protein, iron, and other nutrients.

- Look for nutrition labeling at the meat counter. Single-ingredient raw meat and poultry products, such as ground beef, chicken breasts, and whole turkeys, may be labeled voluntarily. Nutrition labeling is mandatory on most processed meat and poultry. When shopping for processed meats, such as hot dogs and deli meat, use labels to find leaner choices.

- Find other label clues, too. For beef, the terms *loin* and *round* suggest lean choices. On pork, lamb, and veal, your hints are the words *loin* and *leg*.

- For a change of taste, enjoy venison, rabbit, buffalo meat, and veal (except for veal breast)—all of which are lean, delicious, and novel choices.

AT THE SUPERMARKET

- Turkey isn't just for holidays anymore. Cuts of all kinds are sold as versatile, low-fat cuisine. Look for turkey breast cutlets, tenderloin, 99 percent fat-free ground turkey, and 95 percent fat-free turkey franks. Besides being an excellent protein source, turkey is low in fat, saturated fat, and cholesterol. Three ounces of cooked, skinless turkey breast have just 1 gram of fat, 55 milligrams of cholesterol, and 120 calories.

- Unless you're roasting a whole bird, choose light-meat portions of chicken or turkey, such as the breast. Light meat is much leaner than dark meat.

- Buying a whole turkey? Avoid self-basting or butter-injected birds, injected with fat to keep them moist. Instead buy a fresh turkey to baste with fruit juice or broth.

- Buy organ meats—liver, heart, kidneys, sweetbreads, tongue, tripe, brains—only occasionally. Their fat content varies, and organ meats have significantly more cholesterol than lean meat.

AT THE SUPERMARKET

12

PLAN FISHY MEALS.

On the whole, fish, especially the white-flesh varieties, have much less fat than other protein-rich foods. However, their low fat content is just part of the benefit. A large proportion of the fat that is in seafood is polyunsaturated, not saturated.

Fattier fish have the omega-3 advantage as a good source of these polyunsaturated fatty acids. How do they contribute to your health? Research suggests that omega-3 fatty acids can help reduce the risk of heart disease by decreasing the tendency of blood platelets (the part of blood that initiates the clotting process) to stick together and by lowering blood triglyceride levels.

By enjoying seafood at least twice a week, you not only get the omega-3 fatty acids in your diet, but you also replace other, more fat-laden dishes. Here are some tips to help you enjoy the bounty of the sea:

- Know the qualities of fresh fish before you buy. For fin fish, look for bulging, clear-looking eyes, bright red gills, scales that cling to bright and shiny skin, a mild and fresh aroma, stiff fins, and firm flesh. Fresh shellfish should not have a strong odor.

- To estimate fat content, look at the color. Firm, dark-colored fish, such as salmon, swordfish, and mackerel, contain considerably more fat than white fish, such as flounder, halibut, and cod, which are virtually fat free. (Note: To keep the fatty fish even leaner, remove the skin and visible fat the way you would with meat and poultry.)

- Preparing shellfish? Use mostly clams, crab, and scallops, rather than shrimp or lobster. They're all low in fat if you avoid cooking them with rich sauces or butter. However, 3 ounces of raw shrimp supply 130 milligrams of cholesterol. The same amount of clams has a mere 29 milligrams of cholesterol.

- Buying canned salmon, sardines, tuna, or other fish? Choose water-packed varieties. When canned in oil, the fat content multiplies as much as seven times.

13

REMEMBER THE FIVE-A-DAY ADVANTAGE.

An apple a day is not enough! Health experts recommend at least five fruit and vegetable servings each day for better health. Besides their good taste, these foods provide vitamins A and C, complex carbohydrates, and fiber. They help reduce the risk of heart disease and certain cancers. They have little or no fat, and they're cholesterol free!

Filling up on fruits and vegetables is one strategy for cutting back on fat and cholesterol. When you include enough fruits and vegetables in your diet, you may eat fewer high-fat foods. Fruit can be particularly helpful in controlling fat in your diet. Including enough fruit in your daily routine can help you cut down on your cravings for high-fat sweets. Fruit is a naturally sweet treat. Next time you get a craving, have a peach or a pear instead of a slice of cake.

Here are some hints to help you find room for and get the most benefit from fruit:

- Keep a bowl of fresh fruit on the counter or in the refrigerator, so it's easier to reach for than high-fat snack food.

AT THE SUPERMARKET

- Try something new for more variety! Experiment with tropical fruits on breakfast cereal, fruit salads, and frozen yogurt: Guava, mango, papaya, and carambola (star fruit) are all delicious and exotic.

- Make a vitamin C–rich fruit your breakfast tradition. If juice is your choice, make it 100 percent juice. Either way, it has absolutely no fat, and it's a great way to start the day.

- Go easy on avocados and guacamole, which is made from avocado. They're high in fat—15 grams in half an avocado.

- Instead of rich fruit pies and pastries, enjoy fresh fruit for dessert. For a warm dessert, bake apples or poach pears in grape juice or wine; sprinkle granola on top for extra crunch and flavor.

Here are some hints to help you get more out of your vegetables:

- To keep their low-fat quality, microwave, simmer, steam, or stir-fry vegetables. Don't deep-fry them or serve them in heavy sauces. Cook just until they are tender-crisp. That way, they'll retain their nutrient value, color, flavor, and appeal.

- Increase the recipe yield by adding extra vegetables to pasta sauces and casseroles. For example, add mushrooms, bell peppers, and even broccoli to your pasta sauce.

- For a change of pace and a low-fat meal, make vegetables "center plate." Stuff a baked potato with chopped steamed vegetables and shrimp; chili beans and onions; or peppers, salsa, and shredded cheese. Vegetables can make a delicious and filling main dish.

- Create "signature" pizzas with a variety of fresh, thinly sliced veggies from artichoke hearts to sliced zucchini. Broccoli florets, shredded carrots, green onions, tricolored bell peppers, mushroom slices, and chopped tomato are all great toppings for a pizza. Use the premade pizza shells or layer combinations of these same vegetables in a baked vegetarian lasagna.

- Toss pasta with a combination of cooked, tender-crisp vegetables for your own pasta primavera. Add a touch of olive oil, herbs (basil and oregano), and fresh Parmesan cheese, too.

AT THE SUPERMARKET

14

GO FOR GRAINS.

Contrary to popular belief, carbohydrate-rich foods such as bread, cereals, rice, and pasta aren't fattening—unless you smother them with high-fat sauces and spreads. What's more, they make great extenders, stretching a recipe so that small portions of high-fat ingredients, such as meat, go further. In fact, grains are the foundation of a healthful diet; and although 6 to 11 servings per day may sound like a lot, most people need to eat more than they do.

Try these ideas for putting grain products in meals and snacks:

- Feature rice or pasta as the main food in a meal. Combine them with smaller portions of meat, poultry, fish, beans, vegetables, cheese, and low-fat sauces. For example, toss your favorite pasta with diced chicken, vegetables, and a tomato-based pasta sauce.

- Experiment with grains that may be new to you: barley in soup, bulgur with fresh vegetables in salads, and couscous as a pasta or rice alternative.

- Use Nutrition Facts on food labels to choose breads with less fat. On average, bagels, bread sticks, English muffins, French bread, pita pocket bread, and tortillas have 3 grams of fat or less per serving. Cornbread, muffins, and scones have somewhat more, about 5 grams of fat per serving. Croissants, sweet rolls, and doughnuts usually have the most, with at least 10 grams of fat and often considerably more. Look out for the size, too. The large muffins and scones popular in some places can really blow your fat budget.

- Go easy on sauce; enjoy pasta the way Italians do. Add variety by using pasta with different shapes and flavors. Add fiber with whole-grain pasta. (For lower-fat sauce ideas, see tip 20.)

- When you get the urge to crunch, choose low-fat crackers, such as melba toast, saltines, and rice crackers, as your snack rather than potato chips. Check the Nutrition Facts panel for crackers that match your fat budget. (Remember to check the ingredient list for hydrogenated fats and tropical oils.)

AT THE SUPERMARKET

15

MAKE WISE DAIRY DECISIONS.

Calcium is a mineral vital to the health of your bones and teeth. Women, in particular, need it to stave off the debilitating effects of the bone-thinning disease osteoporosis later in life. Calcium also plays a major role in proper muscle contraction, including your heartbeat, and it may help protect against high blood pressure, too.

Dairy foods (milk, yogurt, and cheese) are the body's best source of calcium. The number of servings you need daily from the Milk, Yogurt, and Cheese Group depends on your age. Children need at least three servings so they get enough calcium for their growing bones; teens and young adults to age 25 need four servings; and older adults need at least two. Mounting evidence suggests that adults (especially menopausal women, who lose bone mass rapidly) may benefit from even more.

Along with its much needed calcium, dairy products can also be loaded with fat—and saturated fat, at that. So, how do you get enough calcium and other dairy benefits

AT THE SUPERMARKET

without all the fat? Make the right dairy choices.

No matter what type of milk you choose—skim, low-fat, or whole—one 8-ounce serving (one cup) supplies over 30 percent of the calcium the average adult needs in a day. A cup of yogurt—again, nonfat, low-fat, or full-fat—has about the same amount of calcium as a cup of milk. To cut back on fat, choose the lower-fat varieties of these same dairy foods. You've got nothing to lose but the fat. (Skim and low-fat milks are not advised for children aged two and under who need the calories and essential fatty acids for normal growth.)

Buttermilk is another low-fat dairy choice. Enjoy its thick consistency and tangy taste just as it is! Or whirl buttermilk in a blender with fruit for a quick, low-fat shake.

Cheese is a good source of both protein and calcium. You can control the fat from cheese with smaller portions. Grate or shred the cheese; you'll probably use less than you would with a slice or chunk of cheese. Look for lower-fat varieties, such as part-skim mozzarella.

As a sandwich or bagel spread, blend low-fat ricotta or cottage cheese in a food proces-

AT THE SUPERMARKET

sor until it's thick and smooth. Add herbs for extra flavor. Also, look for low-fat or fat-free cream cheese. Regular cream cheese gets most of its calories from fat. The same spread can be a good substitute for mayonnaise on sandwiches or in salads.

Boost calcium in casseroles, mashed potatoes, meat loaf, and thick soups without adding fat, by blending in nonfat dry milk powder. No one will know the difference.

For a dairy dessert, enjoy frozen yogurt; it has a lot less fat than ice cream. You might also try ice milk. If you can't resist premium brands of ice cream, save fat elsewhere. Perhaps drink only skim milk so you can enjoy a scoop or two later!

If you like adding cream to your morning coffee, try evaporated skim milk. It has a thick consistency, yet it's almost fat-free. Or, slowly cut back on the amount of cream you use, and begin substituting whole milk and then even one or two percent milk. If you do it gradually, you may find it easier to adjust. Don't be fooled by nondairy powdered creamers. Although cholesterol free, they are often made with saturated coconut or palm oil. Remember to read the ingredient list (see tip 10).

AT THE SUPERMARKET

16

TRY MODIFIED VERSIONS OF TRADITIONAL FAVORITES.

Taste, nutrition, price, and product safety—these top four factors most influence food shopping decisions today. To meet the demands of health-conscious consumers, food manufacturers offer more choices now than ever before. Many manufacturers have changed the makeup of their prepared foods, giving you one more strategy for cutting your fat.

Today's supermarkets sell new versions alongside many traditional foods. Fat-free salad dressing, no-cholesterol egg substitutes, low-sodium peanut butter, high-fiber cereals, 90 percent fat-free franks, and reduced-fat cheeses are among hundreds and hundreds of new foods on supermarket shelves. Modified foods have a place in a healthful diet. Count them as servings from the Food Guide Pyramid, as you would for traditional foods, and you may come in under your fat budget.

Use nutrition content claims to spot foods with less fat, saturated fat, or cholesterol in

AT THE SUPERMARKET

almost every department of the supermarket. For specifics, study the Nutrition Facts panel and the ingredient list. (For specific definitions, see tip 9.) How does the modified food compare with its traditional counterpart? Even if it's cholesterol free, for example, it may not be calorie free or fat free.

Although one substitution will not drastically change your fat intake, in combination, these "lite" products give you plenty of ways to whittle down your total fat, saturated fat, and cholesterol intake. Here are some specific ideas:

- If you're a regular sandwich eater and prefer mayonnaise as a spread, don't hold the "mayo." Instead use fat-free mayonnaise, and save 11 fat grams per tablespoon. Or use the low-fat mayonnaise blended with Dijon mustard to make a little go a long way.

- Make frittatas, quiches, and scrambled eggs with cholesterol-free egg substitute.

- Top your baked potato with fat-free sour cream or salsa.

IN YOUR KITCHEN

17

TRIM THE FAT.

It may seem like an obvious way to cut your fat intake, but many people don't think of it or don't take the time to do it. Trim most of the visible outside fat layer from meat before you cook it. Simply removing that layer can slash the meat's fat content. Removing the extra fat and grease can also make the dish more attractive.

You may be surprised by the fat savings on a 3½-ounce broiled steak. A choice grade sirloin, with the fat trimmed to one-quarter inch has 17 grams of fat and 270 calories; the same meat with all the fat trimmed has 12 grams of fat and 230 calories.

Any fat that you can remove before cooking reduces the fat content of the meat. Making meat loaf, hamburgers, or chili? Use extra lean ground meat and completely drain the grease.

IN YOUR KITCHEN

18

GO SKINLESS.

Skin the poultry, cut the fat! Poultry is among the leanest sources of protein around. Much of its fat is unsaturated. Just under the skin, there's a layer of fat. Remove it, and cut fat by more than 50 percent. As with meat, trim any other visible fat. Fat often lurks inside the cavity, especially near the opening.

To keep poultry tender and moist, cook it with the skin on, then remove it just before serving. You'll trim almost as much fat as with skinless cooking, but you'll end up with a tastier dish, because skin insulates the poultry. Loosen the skin by running your fingers between the flesh and skin. For extra flavor, stuff seasonings under the skin and inside the cavity. Try rosemary, pepper, and a pinch of salt. Or, you can spread some lemon juice, thyme, and dry mustard under the skin and cook it; before serving, remove the poultry skin, and enjoy the flavor. Many other herb, spice, and citrus combinations would also work; do a little experimenting.

IN YOUR KITCHEN

19

DE-FAT YOUR COOKING STYLE.

You can cut down on the fat in your diet without giving up the foods you love. Much of the fat in our diets can be avoided by preparing food in a low-fat style. De-fat your cooking style without giving up great taste by following these suggestions:

- Prepare baking pans with vegetable spray, or use parchment paper. Either way, food won't stick to the pans, and you will have avoided adding extra fat.

- Cook in nonstick pots and pans and in heavy iron skillets. They require much less cooking oil; nonstick surfaces may not require any oil at all. Heavy skillets cook more evenly than thinner pans, so they require less fat to keep food from sticking.

Vegetables

- Simmer, steam, stir-fry, or microwave vegetables. To retain vitamins, cook them quickly in little or no water. Avoid deep-frying, as you might with potato fries and onion rings.

- Sauté vegetables in a little de-fatted broth, tomato juice, wine, or water, rather than oil. Or use just a little vegetable spray in a nonstick pan.

Grains

- Consider the "pasta-bilities." Skip the oil in water used to cook pasta. Using enough water—5 quarts per pound of pasta—and stirring vigorously eliminates the need to add oil to prevent sticking.

- Cooking rice, barley, or couscous? Boost the flavor, not the fat. Instead of oil, margarine, or butter, add flavor by tossing herbs in the cooking water. Or cook them in de-fatted broth instead of water (see tip 21).

Meats

- Cooking meat, poultry, or fish? Grill, broil, bake, or roast them; save frying for rare occasions.

- Roast or broil meat or poultry on a rack so that it won't absorb its own fat. Instead, fat will drain away. Baste with de-fatted broth, tomato juice, fruit juice, or wine rather than the drippings.

- When you need to brown meat, broil it on a broiler pan—one that allows the fat to drip away. For meat sauces, chili, and

IN YOUR KITCHEN

sloppy joes, microwave ground meat in a microwave-safe strainer placed over a bowl; fat drains as the meat cooks.

- Cook "under cover." Braising, or poaching, is another way to cook poultry, fish, and meat without adding fat. Braising and poaching mean covering with liquid, then cooking slowly, often in a covered pot. For delicate flavor and tenderizing action, use fruit juice or wine rather than water or broth as the liquid, and add a bay leaf, thyme sprigs, and an onion stuck with cloves to braise with the meat.

- Instead of breading and frying fish, veal cutlets, and chicken breasts, top them with a seasoned crumb mixture, moisten with a little oil, and bake. Breading acts like a blotter that absorbs fat, keeping fried food loaded with fat. The baked alternative allows fat to drain away.

- To seal in the juices and flavor, cook poultry and fish in parchment or foil. Before you wrap it up, add sliced raw vegetables or fruit, such as carrots, zucchini, onions, lemons, or prunes, for extra flavor. Then bake the whole pouch in the oven or on the grill.

20

USE A LIGHT HAND WITH SAUCES AND SUCH.

Onion dip, creamy salad dressing, Hollandaise sauce—traditionally, many dips, dressings, sauces, and spreads have pumped up the fat and calories in our favorite meals and snacks. With '90s-style cooking, you can still enjoy the flavor and appeal without overdoing the fat.

Dips

- Give it a little salsa. Enjoy raw veggies and crackers with salsa rather than mayonnaise-based and sour cream dips. Compare: ¼ cup of sour cream dip has about 6 grams of fat and 80 calories; ¼ cup of salsa has 0 grams of fat and 35 calories.

- In dips that call for sour cream, go "50–50." Use half sour cream and half nonfat plain yogurt.

Dressings

- Dress salads simply. Use a very light hand on salad dressings of all kinds.

- Look for low-fat, low-calorie versions of your favorite prepared salad dressings on supermarket shelves.

IN YOUR KITCHEN

- Keep the flavor but lower the fat in your homemade vinaigrette dressing by changing the ratio of oil to vinegar. Rather than three or four parts oil to one part vinegar, reverse the amounts, using just a splash of oil. Then, flavor the dressing with fresh juice and herbs.

- When you crave a creamy dressing, make your own with low-fat plain yogurt. Blend yogurt with fresh herbs, such as tarragon, chives, parsley, or dill, a splash of orange or lemon juice, and perhaps a little milk to thin it out.

Sauces

- Use thick vegetable sauces for your hot pasta dishes. You can use these same low-fat sauces to moisten pasta salads instead of adding all of that mayonnaise.

- Prepare creamy, low-fat sauces and soups, but instead of cream, blend equal amounts of nonfat plain yogurt with low-fat or nonfat cottage cheese.

- When "saucy" recipes call for cream, use evaporated skim milk. It's rich in calcium, thick and creamy, and almost fat free.

IN YOUR KITCHEN

- Instead of rich gravy, skim off the fat from meat and poultry juices. Then, reduce the pan juices to concentrate the flavor.

- Adjust casserole recipes that use cream of mushroom or celery soup as a sauce base. Instead use half soup and half skim milk, and cut fat from 7 grams to 4 grams.

- For a sweet, refreshing dessert sauce on frozen yogurt, fruit, or angel food cake, puree berries. Sweeten the sauce with honey, mint, ginger, or the "sweet" spices—allspice, cinnamon, and nutmeg.

Spreads

- Try fruit chutney as a low-fat spread on toast and sandwiches.

- For a creamy spread, drain plain, low-fat yogurt by setting it in a fine sieve, cheese cloth, or coffee filter overnight. Blend it with chopped salmon, chives, and other herbs for a delicious, cream cheese–like spread. (It's great on bagels!)

- Which margarine? Choose tub varieties rather than stick margarine. Although the total fat content remains about the same, you can cut saturated fat from 4 grams to 2 grams per tablespoon.

IN YOUR KITCHEN

21

DE-GREASE SOUPS AND GRAVIES.

Being lighter than water, fat rises to the top of soups, stews, gravies, and pan juices making it easier to remove and discard. Each tablespoon of fat skimmed off takes away 13 grams of fat and 117 calories from the finished dish! Try these de-greasing methods:

- Make homemade sauces, soups, and stews ahead, then refrigerate them. When chilled, any fat hardens on top. Just lift it off with a spoon.

- De-fat canned broth, too. Refrigerate it before opening, just as you would homemade broth.

- In a hurry? Add ice cubes to broth to make fat rise.

- For pan juices from meat or poultry, spoon off as much fat as you can, or pour the juices into a fat separator; once fat rises to the top, this type of pitcher allows you to pour out the liquid on the bottom.

IN YOUR KITCHEN

22

ALTER RECIPES TO STRETCH HIGH-FAT INGREDIENTS.

With a little give and take, food experts in test kitchens create great new recipes. Adjust your favorite recipes, too. Cut down on fat—don't cut it out—and enjoy the results just as much.

- Make a flavor shift in homemade muffins, breads, cakes, and cookies. Instead of nuts and chocolate chips, add raisins, apricots, prunes, and dates for flavor and texture but almost no fat.

- Use smaller amounts of high-fat ingredients. If a recipe calls for a ½ cup of pecans, use a ¼ cup and slice the fat from nuts in half.

- In cakes, brownies, and breads, replace half the oil, margarine, or butter with fruit puree, such as thick applesauce or pureed ripe bananas. Purees tend to work better for items that are supposed to have a moist, heavy texture such as quick breads and apple spice cake.

IN YOUR KITCHEN

23

LOSE FAT, NOT FLAVOR.

What makes fat so important in the kitchen? For one, fat carries flavor. It also helps make foods tender and moist. With less fat in food preparation, other flavorful ingredients and preparation techniques can take up the slack.

- For a spark of flavor, add flavored vinegars to your salads and sauces and to soups such as lentil, black bean, and gazpacho. Herbed vinegars are easy to make. Simply add your favorite combination of fresh herbs, peppercorns, and perhaps garlic cloves to cider, white, or wine vinegar in a sterile bottle. Seal and store for at least one week in a cool, dark place.

- Marinating enhances the natural flavor of food and acts like a tenderizer. Make marinades with little or no oil. Tenderizing comes from the acidic ingredients, such as citrus juice and vinegar. Add some sweet fruit juice, honey, or pureed fruit to balance the acidity. Fresh herbs add extra flavor to the marinade.

- In casseroles, salads, and rice and pasta dishes, use strong-flavored cheese, such

IN YOUR KITCHEN

as Romano, Parmesan, feta, or blue cheese. Although their fat content is similar to other whole milk cheeses, they have a lot of flavor; a small amount goes a long way.

- Bring out the flavor of food with fresh, rather than dried, herbs when you can get them. Fresh herbs are more flavorful than dried. The most common varieties—parsley, chives, oregano, and basil—offer great flavor on their own or blended together on pasta, rice, poultry, and fish. If you can't get fresh herbs, there are plenty of good dried herbs. Find a reliable store with quality seasonings.

- When you do want to enjoy the flavor that butter imparts to vegetables, rice, and pasta, add a small amount right before serving. Cooking tends to dilute the flavor, leaving you with the fat but not the taste.

- Flavor carrots, peas, asparagus, and other vegetables with sweetness by tossing them with honey, fruit concentrates, or brown sugar. Mint and ginger also offer a nice flavor complement.

- Add a burst of flavor to meat, poultry, fish, sandwiches, and other dishes with

IN YOUR KITCHEN

low-fat condiments such as mustard, horseradish, salsa, and chutney.

- Harness the flavor power of chilies from mild to very hot! The oil in chilies lingers with flavor. Although we associate chilies with Mexican and Thai foods, a dash of hot pepper sauce adds spark to any cuisine. Try adding some to pasta sauce.

- Rather than adding a creamy sauce, intensify the natural flavor of fresh veggies by roasting them at high heat or grilling them over hot coals. Brush on a little olive oil for extra flavor.

- Follow the seasons. The best flavor of all comes from high-quality foods served at their peak of freshness. Many fresh fruits and tomatoes taste best in season, when they ripen on the tree or in the field.

- Store food properly to keep it at peak quality and flavor. Basic rules apply: 1) keep foods in well-sealed containers so they avoid absorbing the odors of other foods; 2) refrigerate perishable foods; 3) store nonperishable foods in a cool, dark place away from household chemicals; and 4) freeze foods in tightly wrapped packaging to avoid freezer burn, and use while the food remains at peak quality.

24

MAKE EASY SUBSTITUTIONS.

If you want to make fat trade-offs, you can adjust your recipes with these easy substitutions. These small changes often go unnoticed. However, you may want to experiment a bit, especially with baked goods, to make sure you get the texture right.

When the recipe calls for...	Use...
whole milk	2%, 1%, or skim milk
sour cream	plain, low-fat yogurt; lite or nonfat sour cream; or finely strained yogurt (see page 65)
cream	evaporated skim milk
premium ice cream	ice milk, frozen yogurt, sorbet
regular ground beef	extra lean ground beef or lean ground turkey
1 whole egg	2 egg whites, ¼ cup of egg substitute, or 1 egg white plus 2 teaspoons of oil
1 egg (for thickening)	1 tablespoon of flour
creamed canned soups	½ can of soup and ½ can of evaporated skim milk
nuts	dried fruit, such as raisins
1 ounce of baking chocolate	3 tablespoons of cocoa powder and 1 tablespoon of oil
mayonnaise	low-fat or fat-free mayonnaise or plain, low-fat yogurt

IN YOUR KITCHEN

25

WATCH PORTION SIZE.

What's on your plate? Chances are, you underestimate the size of your portions, then overspend your fat budget. Your fat and cholesterol intake depends not only on what you eat, but also how much!

Many people consider meat the center of the dinner plate. You don't need an 8-ounce steak. Cutting the portion in half will split the fat and cholesterol content down the middle. Smaller servings—3-ounce, cooked and boneless portions of meat, poultry, and fish—can provide all that you need in one meal. After all, a total of only 5 to 7 ounces per day is the current recommendation. Because of cooking losses, 4 ounces of raw meat equal about 3 ounces when cooked.

How much is a 3-ounce portion? For beef, pork, and fish, it's about the size of a deck of cards. On average, a 3-ounce portion of cooked poultry on the bone is the amount from half of a chicken breast, one chicken thigh, two drumsticks, or half of a Cornish game hen.

IN YOUR KITCHEN

Get a set of measuring cups to help you judge portions. Serve your meals and snacks on smaller plates and bowls. That way, a standard portion looks like more.

Still hungry? Build in portion control for higher-fat entrees and desserts. Fill up on low-fat side dishes such as grains, salads, fruits, and vegetables.

When you eat out, watch out for these restaurant traps:

- Most restaurants serve 6- to 8-ounce portions. Order the smallest steak. Split an entree with a friend, and order an extra salad and vegetable side dish. Or, eat only what you think is 3 ounces and take the rest home in a doggy bag for another meal.

- All-you-can-eat buffets and salad bars can throw portions out of control, too. Take the small plate, and avoid going back for seconds.

- Go for the small order of fries and other high-fat foods. A large order of fries can have about 400 calories and 22 grams of fat. Although still high, the small order has 220 calories and 12 grams of fat.

26

BE "EGGS-TRA" SMART.

Say "cholesterol" and most people think "eggs." Eggs are an economical source of protein, vitamin A, and iron, but they also supply more cholesterol than most other foods. Egg yolks, not whites, contain cholesterol. In fact, the yolk of one large egg has about 215 milligrams of cholesterol, compared with no cholesterol in the egg white. Even though dietary cholesterol intake is not the greatest concern when it comes to blood cholesterol, on average, it's smart to consume no more than 300 milligrams of cholesterol a day.

To control dietary cholesterol, health experts recommend eating no more than four egg yolks a week, including those used as ingredients in baked foods and other dishes. How do you make "eggs-tra" smart choices? In recipes that call for whole eggs, such as breads, pancakes, puddings, and many casseroles, you have some alternatives:

- Try the two-egg-white solution. For a cholesterol-free ingredient, substitute two egg whites for one whole egg.

IN YOUR KITCHEN

- With egg whites only, scrambled eggs and other dishes may seem pale and bland. To retain some of the color and flavor, substitute one whole egg and two egg whites for two whole eggs. You'll still cut the cholesterol content in half.

- Replace whole eggs with egg substitutes. Usually, a ¼ cup of egg substitute equals one whole egg. However, check the substitution guide on the package. (Many egg substitutes are frozen. Allow thawing time when you use them.)

Do egg substitutes have less fat? Check the label. To help egg substitutes perform like whole eggs, manufacturers may add oil to replace the fat in the yolk, along with sodium for flavor and yellow coloring. Egg substitutes work best in baked goods, batters, and mixed dishes.

As another alternative, look for reduced-cholesterol, liquid whole eggs in your grocery store. Their natural color and flavor come from the yolks. However, processing removes most of the fat and cholesterol. They work in any recipe that calls for whole eggs. (See tip 24 for other substitution ideas.)

SNACKS AND MEALS

27

CONTROL YOUR "FAT TOOTH."

Do you have a sweet tooth or fat tooth? Some evidence suggests that a hunger for fat, not sugar, may be the real temptation of rich desserts. Your craving for chocolate, cake, a thick shake, or a doughnut may have nothing to do with its sweetness, but rather its fat content.

What makes high-fat foods so appealing? Fat makes pastries flaky and tender, ice cream smooth and creamy, chocolate rich, and meat juicy. Its unique texture and taste make it hard to replace. To many people, the high-fat, high-sugar combo tastes best of all.

The next time you have the urge to splurge, redirect your fat tooth:

- Share a slice of chocolate cake with a friend and cut the fat in half.

- For a thick, creamy shake, puree low-fat yogurt with fruit.

- Crunch on cookies with less fat; gingersnaps, vanilla wafers, and animal crackers are good choices.

SNACKS AND MEALS

28

GRAZE ON SMART SNACKS.

Snacking—it's an all-American pastime! Depending on your choices, snacks can work for you. They can provide servings from the Pyramid and nutrients you need without overdoing fat grams or calories. Snacks curb your appetite, too, so you don't overeat at dinnertime. When you get the urge to nibble, try lower-fat snack combos for two food groups in one:

- low-fat yogurt blended with fresh fruit
- salsa on low-fat whole-grain crackers
- raw veggies (broccoli, zucchini, carrots, bell peppers) and low-fat yogurt with herbs
- rice cakes with lean deli meat
- soft tortilla with black bean salsa

SNACKS AND MEALS

29

BOOST YOUR FIBER FACTOR.

Often called *roughage,* dietary fiber refers to the parts of grains, vegetables, and fruits that the human body cannot digest. That's its secret. Instead of being digested for energy or raw materials, fiber moves through your digestive tract, absorbing water, providing bulk, and offering protection from some diseases such as cardiovascular disease and cancer.

Because they provide bulk and often very few calories, high-fiber foods help to fill you up—pushing away extra amounts of higher-fat foods. Many high-fiber foods are also high in complex carbohydrates, but low in fat and saturated fat. Since they come from plant sources, high-fiber foods have no cholesterol unless animal products are mixed with them.

There are basically two forms of dietary fiber—soluble and insoluble—and both have health benefits. The fiber in most foods is a mixture of both kinds, but some foods have more soluble fiber, and others, more insoluble fiber.

SNACKS AND MEALS

Soluble fibers have a unique ability to hold water. They also appear to reduce total blood cholesterol levels. (The big to-do over oat bran a few years ago was based on the soluble fiber it contains.) Oats, legumes (beans, peas, lentils), barley, apples, and citrus fruit all supply soluble fiber.

Insoluble fiber helps move food and waste through your digestive tract. Because it shortens the time that cancer-causing compounds are in contact with the intestines, it may help lower the risk of colon cancer. Insoluble fiber comes mainly from bran, whole grains, vegetables, and fruits with edible seeds.

Boost the fiber in your food choices:

- Keep the skin on! Edible peels on fruits are great fiber sources.

- Eat whole fruits and vegetables instead of drinking the juice. Juicing removes most of the fiber, which is found mainly in the skin and pulp.

- Switch to whole-grain breads, pasta, and breakfast cereals. Sprinkle bran over cereal, fruit, and yogurt.

- Enjoy legumes, such as kidney beans, several times a week.

SNACKS AND MEALS

30

GO VEGETARIAN EVERY NOW AND THEN.

Most of us tend to plan meals around meat, poultry, and fish, but these aren't the only sources of high-quality protein. When vegetable proteins from legumes, grains, nuts, and seeds are teamed up, they too provide high-quality protein without cholesterol and with little or no saturated fat. Depending on the preparation, side dishes, and desserts, vegetarian meals tend to be high in complex carbohydrates and fiber and low in total fat.

The total fat in these foods varies. Legumes have almost no fat, whereas nuts and seeds have relatively large amounts: 16 grams of fat in 2 tablespoons of peanut butter, 19 grams of fat in just a quarter of a cup of seeds, and 23 grams of fat in a third of a cup of nuts. Fortunately, this fat is mainly unsaturated.

Enjoy vegetarian meals one or more times a week. Caution: A strict vegetarian diet with no animal products is not recommended for most people, especially children. Adults who want to follow a strict vegetarian diet that does not contain any animal products at all need to do some extra planning, preferably with the advice of a registered dietitian.

Learn Bean Basics

Many ethnic dishes use beans as a mainstay: Brazilian *feijoada* (black bean and smoked meat stew), Italian pasta e fagioli (pasta and beans), and Puerto Rican rice and beans. For many people, however, bean basics require new cooking techniques:

- Before you cook or soak beans, rinse them to remove damaged beans and any foreign matter.

- To cut down on cooking time, soak beans ahead. Then, let them stand overnight in water: 1 quart of water for 1 pound of beans. If it's warm, keep them in a cool place or in the refrigerator, so they don't start to ferment. Presoaking isn't necessary, but it sure saves time! (Note: From dry to cooked, 1 cup of dried beans equals about 2½ cups cooked.)

- For a quick soak, boil beans in the same amount of water for 2 minutes. Then, cover and set aside for an hour before cooking.

SNACKS AND MEALS

- Drain any soaking water before cooking the beans. Then add fresh water: 4 to 6 cups for every pound of beans (weight before soaking). Cover and simmer until the beans are tender. Depending on the variety, presoaked beans take anywhere from 30 minutes to 2 hours to cook. Check a cookbook for microwave instructions.

- Avoid salting the cooking water. Salt draws the moisture out of the beans, upping the cooking time. If you add salt, sprinkle in a moderate amount toward the end of cooking. You may add other seasonings, however, such as herbs, spices, onion, celery, or ham.

- Add acidic foods, such as tomatoes or vinegar, when beans are almost tender. Acid slows the softening of beans and so increases the cooking time.

- In most recipes, canned beans can replace home-cooked beans. Rinse canned beans before preparing your recipe; the rinse removes much of the sodium added during processing.

EATING OUT

31

CONTROL FAT IN THE FAST-FOOD LANE.

Do you eat in the "fast-food lane?" With our hectic lifestyles, many of us do. Fortunately, there's plenty to choose from. Today's fast-food restaurants supply traditional fare, as well as lower-fat options. Contrary to popular belief, a fast-food meal can be healthful. You just need to know how to order foods that balance your day's choices with variety and moderation.

- Watch out for portion size. A regular 3-ounce burger is an ample portion with about 12 fat grams and 275 calories. The deluxe burger offers a bigger meat serving than you need, and it has, on average, 23 fat grams and 400 calories or more.

- On sandwiches of all kinds, hold the mayonnaise, tartar sauce, and other sauces to cut fat and calories. Use mustard, salsa, relish, and catsup instead, and pile on extra lettuce, tomato, and onion for your "five-a-day advantage."

- Enjoy pizza. It's highly nutritious, representing three or more food groups on the Food Guide Pyramid. Choose toppings

EATING OUT

with little or no fat: Canadian bacon or shrimp instead of bacon, pepperoni, or sausage; chopped vegetables, such as onions, peppers, broccoli, zucchini, and mushrooms. Avoid extra cheese.

- Enjoy roasted chicken instead of fried. If fried chicken is your only option, avoid the "extra crispy" kind. Either way, you can remove the skin to cut down on fat.

- Substitute a side salad with a light dressing for mashed potatoes with gravy. If you prefer the potatoes, hold or get just half the gravy.

- Many people choose salad bars to cut down on fat and calories. However, salad dressings can be a major source of fat. A trip to the salad bar can add up. A 1-ounce ladle of regular dressing has 7 grams of fat and 70 calories. Choose the reduced-fat dressing and don't take too much, or use vinegar and just a little oil. Take a light approach with lots of fresh greens, fruits, and vegetables. Go easy on creamy coleslaw, macaroni salads, and potato salad made with regular mayonnaise. For extra fiber, add beans!

- Instead of fries, enjoy a baked potato. A large order of fries has 22 grams of fat; a

EATING OUT

plain baked potato has less than 1 gram of fat. Top it with chopped, steamed veggies or salsa.

- Drink milk with fast-food meals to boost calcium; many chains offer 2 percent milk, and sometimes you can even get 1 percent milk. For more vitamin C, order fruit juice.

- For dessert, enjoy frozen yogurt. One half-cup serving has 3 grams of fat and 110 calories, compared with 15 grams of fat and 260 calories in a fruit pie. Or enjoy fresh fruit from the salad bar.

- For breakfast, order pancakes or cereal with low-fat milk. Eggs, sausage or bacon, and hash browns is a high-fat combo. Order a bagel or English muffin; they have less fat than biscuits, croissants, or doughnuts.

For more information on the nutrients and calories in fast-food menus, check with the restaurant. Nutrients and calories on similar menu items differ from chain to chain. Some display their nutrition analysis on the menu or on take-home brochures. Others provide the data with a written request.

EATING OUT

32

ORDER WISELY.

Today, restaurants are full of menu choices, many prepared with nutrition-conscious customers in mind. From fine-dining restaurants to casual cafes, cafeterias, and family eateries, you'll find a wide variety of foods that fit within the Food Guide Pyramid guidelines. Menus offer more and more choices with fewer calories and less fat, cholesterol, and salt.

When you eat out, control the food on your plate. Study the menu. Ask questions before placing your order. Select foods to match your own fat budget. If it isn't served as you requested, you can send it back.

- *Ask about serving sizes.* Can you have a "petite" rather than an 8- or 12-ounce steak, a half portion or appetizer portion for the main course, or one order to split with a friend? If the portion served is truly bigger than you need, ask for a take-home bag.

- *Ask how foods are prepared.* What ingredients are used? How are foods cooked—fried or with sauces, or are they broiled, roasted, steamed, or baked? Is the meat well trimmed? Can you order lean meats, such as London broil, filet mignon, or flank steak?

- *Make your own requests.* Have salad dressing, sauce, or sour cream served on the side (so you control the amount). Have your vegetables steamed without sauce, margarine, or butter. Ask for fresh lemon to squeeze as a flavor enhancer for broiled fish or chicken.

- *Request items that aren't on the menu.* Can you order: fresh fruit for dessert? raw veggies for an appetizer? low-fat or skim milk?

- *Remove higher-fat snacks from the table.* If you can't resist the urge to nibble, have the temptation taken away. That includes many dry snacks, such as chips, buttery crackers, and nuts.

When you fly, you have more menu choices than you may realize. If you order a day or more in advance of flight time, you can ask for a special meal, including low-calorie, low-cholesterol, or vegetarian meals. As an option on the road, take fresh fruit or canned fruit juice in your briefcase; don't rely on the high-fat snacks from vending machines.

EATING OUT

33

LOOK FOR MENU CLUES.

Does the menu have "spa cuisine" or items listed as "on the lighter side"? Claims about nutrition, such as "lean," "light," and "low cholesterol," appear on some menus. However, they are not regulated like food labels are. Instead, their meanings depend on the chef.

In addition to asking the right questions and making the right decisions, it pays to be savvy in the way food is presented. Many people think that if it's listed under the salad section, it's got to be low fat. Pay attention to what the menu really says:

Menu Clues That Signal...

Higher Fat	Lower Fat
au gratin	steamed
creamed *or* in cream sauce	au jus *or* in its own juice
breaded	roasted
hollandaise *or* bernaise *or* bechamel	broiled
fried, french-fried, batter-fried, *or* pan-fried with gravy *or* pan gravy	poached
	stir-fried
scalloped *or* escalloped in a pastry shell	grilled

88   50 WAYS TO LOWER YOUR FAT & CHOLESTEROL

EATING OUT

34

GIVE YOUR TASTE BUDS A PASSPORT.

When looking for a restaurant, you want to find a place that has good food and offers some variety. You don't always want to go to the same place and order the same low-fat option—and you don't have to. Many ethnic cuisines offer variety and taste that is traditionally low in fat.

First Stop: Italian Fare

Everybody seems to like spaghetti and pizza. Today, we've expanded our experience with Italian cuisine. Pasta dishes appear on all kinds of restaurant menus as filling, nutritious, and economical dishes. Enjoy pasta of all kinds. Restaurants have varieties and shapes that you may not find in most supermarkets. Here are some ways to enjoy lean Italian food:

- If portions seem too big, order the appetizer portion of pasta for your entree.

- Pick your sauces wisely. Red sauces, made with tomato, and broth-based sauces tend to be lower in fat than creamy white sauces. Try pasta primavera with a light touch of oil and fresh vegetables and herbs.

- Go easy on garlic bread, which tends to be broiled with high-fat spreads. Ask for plain Italian bread with olive oil, margarine, or butter on the side. Seasoned olive oil is served in some Italian restaurants as a dipping sauce for bread. Dip lightly; even though it's high in monounsaturated fatty acids, olive oil still contributes to your total fat intake.

- Balance your meal with a fresh, crisp salad, lightly dressed with vinegar and oil, or have minestrone soup made with a variety of vegetables.

- Besides pasta, explore risotto and polenta as other strategies for getting servings from the bread group.

- Be aware that two popular dishes are often breaded and fried and can be very high in fat: veal and chicken Parmigiana. Veal scaloppine is often served in a high-fat sauce. Order veal or chicken marsala, cacciatore, or piccata more often.

- Enjoy a refreshing Italian fruit ice for dessert. It's fat-free.

Next Stop: Asian Cuisine

Chinese, Japanese, Thai, Vietnamese, and many other Asian cuisines, place great emphasis on rice, noodles, and a wide vari-

EATING OUT

ety of vegetables. These foods extend meat, poultry, and fish, making a 3-ounce portion from the meat group look much bigger!

- Choose stir-fried dishes; go easy on deep-fried. Many foods that start the menu (egg rolls, spring rolls, wontons, and crab rangoon) are fried. If you choose one fried dish to share at a meal, balance your choices with others that are simmered, roasted, steamed, or broiled. In Japanese restaurants, the term *yakimono* refers to broiled; *tempura* is fried; and *teriyaki* is broiled in a sauce.

- Look for steamed dishes on the menu: for example, steamed rice, steamed vegetables, and steamed dumplings.

- Enjoy broth-based soups, such as miso, wonton, and hot-and-sour soups.

- Enjoy a big bowl of rice as the Chinese do. You'll boost the proportion of complex carbohydrates in your diet.

- Rice isn't the only carbohydrate-rich food in the Asian diet. Noodles are just as popular. As a switch from rice, choose dishes that feature noodles. Chinese *lo mein* and Thai *lad nar* are dishes that use rice noodles; Japanese cuisine sometimes uses buckwheat noodles.

EATING OUT

- Try some vegetarian combinations, such as bean curd (tofu) dishes. In an Indian restaurant, try dishes made with lentils, legumes, vegetables, and rice.

- Plan to take home a Chinese doggie bag. In many restaurants, one dish could serve two customers. Even though it's nutritious, the whole dish may be more food than you need. For portion control, save the extra for a healthful lunch or dinner the next day.

- Top off your meal with a fortune cookie. One cookie has just 15 calories and no fat, and it just might have some good news.

And Next Stop: Mexican Food

Mexican and Tex-Mex foods have become so popular that they're almost all-American! The menu has variety from every food group. Here are some ways to moderate calories and fat:

- Ask for foods served in soft tacos (plain tortillas). A basket of warm tortillas makes a nice meal accompaniment and a great holder for fajitas. Go easy on tortilla chips, though; crisp tacos and chips are deep fried.

EATING OUT

- Add flavor to all kinds of dishes with plenty of salsa. Made with vegetables, it adds to your day's intake of veggies, and it's usually fat-free. Gazpacho, a cold soup made with tomatoes and other vegetables, has little or no fat.

- On tacos, burritos, taco salads, and fajitas, ask for guacamole and sour cream on the side. That way you control the amount. Guacamole is made from avocado, a fruit that is high in fat.

- Look for the lower-fat entrees. Fajitas, for example, are stir-fried chicken or beef with vegetables, tucked in a soft tortilla. Burritos are usually made in a soft tortilla, too. Go easy on chimichangas; the ingredients are about the same as fajitas, except they're deep-fried. Chilies rellenos are often deep-fried, too.

- As an occasional vegetarian meal, have bean burritos with Mexican rice on the side. The combination provides complete protein. If the menu is more extensive, try black bean soup, rice and beans, or black bean salsa.

AN ACTIVE LIFESTYLE

35

GET MOVING!

Estimates suggest that a lack of physical exercise may be a factor in 250,000 deaths per year in the United States. Only 22 percent of American adults get enough exercise to benefit their health. Exercise doesn't need to be strenuous to make a difference. Simply shifting your lifestyle from sedentary to moderately active offers significant health benefits. If the proper incentive is what you're waiting for, here are ten great reasons to exercise:

1. Exercise helps control your body weight. Being trim makes you look and feel good. A brisk daily 30-minute walk burns up to 180 calories, or more than 1,200 calories in a week! That equals about one third of a pound of body fat.

Exercise may offer another weight-control bonus. It may speed up your metabolic rate, so you burn more calories for up to 12 hours after a workout. With regular exercise, you lose weight without feeling deprived. You won't need to cut as many calories from food to lose body fat, because you're burning more calories.

AN ACTIVE LIFESTYLE

Weight control, besides making you look and feel good, has benefits for your heart. When you maintain a healthy weight, your heart doesn't have to work as hard to keep up with the body's oxygen demands.

2. Exercise helps control "good" and "bad" cholesterol. Even a small shift from sedentary to moderate exercise helps increase HDL cholesterol levels. Only aerobic exercises, such as jogging, swimming, bicycling, and brisk walking, boost HDL. After one year of regular, brisk walking, HDL cholesterol levels might be up by 25 percent. Anaerobic exercises, such as strength training and sports that require bursts of energy (football and weight lifting), appear to lower LDL cholesterol levels.

3. Exercise keeps your heart and lungs healthy. Like muscles in your arms and legs, your heart needs regular exercise to stay strong and to circulate blood through your arteries and blood vessels. Blood carries nutrients and oxygen to body cells where they're used to produce energy; it also removes waste from the cells. Exercise improves the circulation of oxygen and nutrients by making the pump more efficient.

AN ACTIVE LIFESTYLE

Exercise builds lung capacity, too, improving the lungs' ability to bring in the oxygen needed to produce energy. Together, a healthy heart and lungs make energy production more efficient.

4. Exercise helps with appetite control. Sedentary people may refrain from exercise because they think their appetite will increase. They believe, what's the point of working out if I just eat more? Interestingly, those who sit around tend to eat slightly more. Some studies suggest that light to moderate activity for up to an hour daily suppresses appetite slightly.

5. Regular exercise builds lean muscle, making your body stronger. The obvious benefit is being able to move and lift things more easily. Strength also helps you withstand physical stress, handle many everyday activities, avoid back injury and strain, and get out of dangerous situations more easily.

With age, sedentary people lose more strength than they realize. Although body weight may stay roughly the same from age 20 to 70, sedentary adults lose about 30 percent of their muscle cells, or about 6 to 7 pounds of lean body mass per

AN ACTIVE LIFESTYLE

decade. Strength training can't replace lost muscle cells, but it can build those that remain.

As a boon to weight control, muscle burns more calories than body fat, because muscle tissue is metabolically more active than fat. If, for example, two people have the same weight, the one with more muscles needs more calories to maintain weight than the one with more body fat. Even at rest, muscle burns more calories than fat.

When it comes to appearance, you look better when your muscles are firm, not flabby. In fact, one pound of body fat takes three times more space than one pound of muscle. That's why lean people look thinner at the same weight!

6. Weight-bearing exercises, such as walking and jogging, promote bone health. Among children, teens, and young adults, these exercises help increase bone density. For middle-aged and older adults, these same exercises slow bone loss that comes with aging. Weight-bearing exercise is a frontline defense against osteoporosis, or brittle bone disease, which can cause crippling and life-threatening fractures late in life.

AN ACTIVE LIFESTYLE

7. Exercise builds physical endurance. Strength is part of physical endurance. Your muscles also use oxygen more efficiently with regular exercise. Whether you're involved in sports or a physically demanding job, regular exercise helps you keep up an intense physical pace without tiring too quickly.

8. Exercise keeps your body flexible and agile. Flexibility allows you to stretch, bend, and twist without strain or injury to muscles, ligaments, or tendons. Not only does it feel good, but a flexible body moves with greater ease and grace.

9. Exercise helps to relieve stress and keep you alert. Physical activity relieves tension that may cause tight muscles or aggravate physical ailments. A good workout often helps you forget "stressors" for a while. When your heart pumps more oxygen to your brain and body cells, you feel more energetic and alert, too.

10. Why not? Exercise can be fun and can improve the quality of your whole life! For many people, an active lifestyle opens up a new world of friendships and recreational interests. Try it!

AN ACTIVE LIFESTYLE

36

TEST YOUR EXERCISE QUOTIENT.

How would you judge your "exer-style?" Do you put your body in motion? Or are you the classic couch potato? Consider the following questions and circle your best answer (**A** for always, **S** for sometimes, **N** for never). Do you...

Walk instead of drive or ride?	A	S	N
Participate in sports rather than watch from the sidelines?	A	S	N
Walk before or after work?	A	S	N
Take a walk on your lunch hour rather than sit and talk?	A	S	N
Work out at an exercise class or gym?	A	S	N
Take time for active play with your family several times a week?	A	S	N
Take stretch breaks during the day?	A	S	N
Make exercise rather than television your leisure activity?	A	S	N
Use the stairs rather than the elevator?	A	S	N
Regularly work on household chores that get you moving?	A	S	N
Periodically get up from the computer or desk for a few minutes of exercise?	A	S	N
Take your family on an active vacation?	A	S	N
Walk at a brisk pace in daily activity?	A	S	N

Count the circles in each column. What do your answers suggest about your lifestyle?

50 WAYS TO LOWER YOUR FAT & CHOLESTEROL

AN ACTIVE LIFESTYLE

37

CHOOSE AN ACTIVITY JUST RIGHT FOR YOU.

For many people exercise is an all-or-nothing notion. They think that only strenuous exercise yields benefits. If real workouts don't appeal to them, they just sit back. On the contrary, the best exercise program is the one that matches an individual's lifestyle, physical ability, and personal interest. The activity you enjoy is the one you'll stick with.

The "no pain, no gain" mentality is a myth. Exercise should be pleasant, not a grind. Moderation is okay. A half hour on the stair climber at the health club is not for everyone. In fact, an exercise program that is overly strenuous isn't even preferable. When an exercise plan is too ambitious, it makes the activity a dreaded experience and one that is more likely to be avoided. Choose an activity you can look forward too.

Even a small increase in exercise makes a difference, no matter what your starting point. Every little bit helps. Substantial health benefits can come from a moderately active lifestyle.

AN ACTIVE LIFESTYLE

38

PLAN AN EXERCISE PROGRAM FOR F.I.T.

It can be difficult to figure out what your exercise priorities should be, but as long as you make it F.I.T., it will work for you. An exercise plan that considers Frequency, Intensity, and Time offers the qualities that promote fitness and health.

Frequency: Three times a week is a good baseline goal. If you have more time, add one or two more days. If you miss one or two times, don't give up. Exercise over time makes the difference. Try to be consistent.

Intensity: This is your perceived level of exertion. If you feel you're working hard, you've probably pushed yourself enough. As your exercise gets easier, increase the intensity. Your own comfort zone within your target heart rate is your workout goal. (To figure your target heart rate, see page 102.)

Time: Make about 30 minutes your goal. If you need to, start with less time, and gradually build up, adding two to three minutes each week. The duration of your exercise is the factor that really ups your fat-burning capacity.

AN ACTIVE LIFESTYLE

FIGURING YOUR TARGET HEART RATE

When exercising, you want to get your heart rate into the target zone to receive the most aerobic benefit. Your target is between 60 and 75 percent of your maximum heart rate (the fastest your heart can go). This maximum changes with age; so use this quick reference chart to find your target heart rate:

Age (years)	Target heart rate (beats per minute)
20	120–150
25	117–146
30	114–142
35	111–138
40	108–135
45	105–131
50	102–127
55	99–120
60	93–116
70	90–113

AN ACTIVE LIFESTYLE

39

WALK YOUR WAY TO HEALTH.

Around the neighborhood, through the park, around the track, in long hallways, walk your way to health! Walking is the exercise of choice for almost 70 million Americans. There's good reason.

Walking promotes heart health and lowers blood pressure. It burns about 100 calories per mile. A good walk strengthens many muscle groups in the legs, abdomen, shoulders, and with a little pumping, arms, too. Walking burns calories and promotes loss of body fat. It relieves stress. Compared with jogging, walking is less likely to result in shin splints or other injury; the impact on the feet is at least 50 percent less. The added bonus: Walking costs no more than a good pair of shoes. And, it's convenient!

To make the most of your walk:

- Start with comfortable walking shoes. They should cushion your stride and offer adequate support.

- Walk for time, not pace. According to some researchers, walking a slow mile can trim as much body fat and burn as

AN ACTIVE LIFESTYLE

many calories as a fast mile. Just walk longer to cover the same distance and burn the same number of calories. The advantage of the faster pace is that you maximize the heart-healthy benefits.

- Get into the swing of things. Bend your arms at a 90-degree angle and pump with each stride. It's likely you'll walk faster and burn more calories.

- Walk with good posture—shoulders back, head high. Step heel first, then toes, rolling forward and always pushing off with your toes.

For more cardiovascular benefits, take brisk power walks:

- Start with a warm-up, as with jogging. Twist, stretch, bend, and walk slowly for five to ten minutes. Give special attention to your hamstrings, calves, and shins.

- When you're done, cool down. Stroll for 10 more minutes to give your heart a chance to slow down gradually.

AN ACTIVE LIFESTYLE

40

FIT IN UNSCHEDULED EXERCISE.

Today's lifestyles are more sedentary than ever. Unlike the past, technology (computers, push-button appliances, electronic tools for the home, to name a few) saves us time and effort. However, we use our bodies less and less. To put more activity in your daily life, fit in extra activity with these strategies; some may seem silly, but every little bit helps:

- Use the stairs rather than an elevator or escalator.

- Get physical with housework. Wash the car. Wash the windows. Wash the dog! Even the "push and pull" of vacuuming builds upper-body strength.

- Park at the far end of the parking lot, then walk the distance.

- Get off the train or bus one stop early. Walk the extra distance.

- If you need to go half a mile or less, walk or ride a bike. Don't drive.

AN ACTIVE LIFESTYLE

- If you have a small grocery order, carry the bags of food to the car, rather than push them in a cart.

- Get a cordless phone so you can walk while you talk.

- Work in the garden. Skip the sit-down mower; instead, walk the mower around the yard. Raking the lawn and shoveling snow offer good upper-body exercise, too.

- Turn on lively music, perhaps your rock radio station. You may move faster doing your chores without even knowing it!

- Enjoy television? Instead of a talk show or sitcom, turn on an exercise video and work out, or use a stationary bike, treadmill, or rowing machine as you watch your favorite show.

- Walk your dog. If you're a cat owner, you might walk your feline on a leash, too.

- Don't just watch the kids play in the backyard, go play with them, or take a walk together.

- For an evening out, go dancing instead of sitting at the movies.

AN ACTIVE LIFESTYLE

41

PUT ACTION IN YOUR WORKDAY.

As the workforce moves from heavy manufacturing to service industries, many workers simply sit at a computer or a desk, or stand at a work station all day. Fewer and fewer work in physically demanding jobs. Eye, back, and neck strain, tight muscles, poor circulation, and mental fatigue result from jobs that keep us sedentary at a work station. When it comes to heart disease and other health problems related to inactivity, the health costs are harder to measure, but certainly high.

You can put exercise into your workday without losing productivity. You might even accomplish more—and feel better doing it. Here's how:

- Unless safety and accuracy are issues, move faster on the job. Rather than a slow saunter, walk briskly from office to office.

- Take an exercise break at work, not a coffee break. Instead of spending 176 calories and 11 grams of fat noshing on a

AN ACTIVE LIFESTYLE

doughnut, burn about 90 calories on a brisk 15-minute walk.

- Keep a pair of comfortable walking shoes in your desk or locker at work. Then you're always ready to walk as far and as long as your break allows.

- In cold weather, walk the hallways, go up and down the stairs, or bundle up for a brisk 15-minute walk outdoors. During warm weather, walking outside during breaks is a natural. Take a walk to a nice picnic spot for lunch—10 minutes there and 10 minutes back.

- Start a 100-mile club at your workplace. Measure a 1-mile course around the parking lot and buildings. Have your "club" clock their individual miles. Recognize each 100-mile walker on the bulletin board.

- If you travel on the job, pack a jump rope and a pair of good shoes. No matter where you are, you'll have equipment for a workout.

- If your workplace has an employee exercise facility, use it!

AN ACTIVE LIFESTYLE

42

GIVE YOUR BODY A REGULAR WORKOUT.

While moderate activity offers many health benefits, more intense activity exercises your whole cardiovascular system. Aerobic exercises, such as walking, jogging, swimming, dancing, cross-country skiing, and bicycling, are sustained, rigorous activities that build endurance, burn body fat, and maintain overall fitness.

Aerobic means "with oxygen." That's the difference between sustained activities and those that require spurts of energy. Aerobic activities require oxygen in the muscles to produce energy; anaerobic activities, such as calisthenics, do not. A healthy cardiovascular system, with a healthy heart and healthy lungs, brings more oxygen and improves your aerobic capacity.

When you are ready to work out, choose activities that exercise muscles in different ways. Go with different exercises that, when combined, include all of these aspects of a workout:

- Build endurance with aerobic activities, such as jogging, dancing, and brisk walk-

AN ACTIVE LIFESTYLE

ing, that increase your heart rate to its target zone (see tip 38). This helps you condition your heart and lungs and burn body fat.

- Increase your strength and lean muscle tissue with activities that require you to lift, push, and pull, such as weight training and calisthenics.

- Increase your flexibility and agility with exercises that stretch, bend, and twist. Flexibility training helps you maintain the normal range of motion around a joint. Stretch gently; avoid jerky or bouncy motions.

Set your target at 1,000 calories a week in any activity you like. At this point, exercise has a significant impact on health; HDLs may also go up measurably. That's about 10 miles of walking or jogging, about 2½ hours of swimming, or about 1½ hours of bicycling at 12 miles per hour.

Remember to do your exercises properly to avoid injury and maximize the benefit. You can help insure proper technique by following the three phases of a good workout:

Warm-Up Phase: Warming up for 5 to 10 minutes gets your heart and muscles ready for more intense activity. It's a safety pre-

AN ACTIVE LIFESTYLE

caution for injury. Start with low-intensity aerobics, such as walking. Get your muscles limber with gentle 20 to 30 seconds of stretches. Stretching helps prevent injury to your muscles, tendons, and ligaments by increasing blood circulation and making them more elastic so they won't snap or tear. Gradually pick up your speed. That helps raise your heart rate from a resting level to a conditioning level.

Aerobic Conditioning Phase: The aerobic phase takes 20 to 30 minutes and involves large muscle groups. It's steady, rhythmic, and nonstop, allowing your heart to beat at its target rate (see tip 38). Work out at a comfortable, yet rigorous, level. You should be able to talk, rather than gasp for air. If you haven't been exercising, target your heart rate at the bottom of the range (60 percent). As your workout gets easier, gradually work harder until you reach the top of the range (75 percent).

Cool-Down Phase: For 5 to 10 minutes, cool down as you did in your warm-up. Slowly decrease the intensity and the speed of your aerobic activity to bring your heart rate down to its resting level. Stretch again to prevent muscle stiffness and soreness.

AN ACTIVE LIFESTYLE

43

MAKE EXERCISE A FAMILY AFFAIR.

Exercise isn't just for you. It should be a family affair! People of every age need exercise to keep fit. When the entire family is involved, exercise is more fun and you are all more likely to stick with it.

- After work, take a walk around your neighborhood together. Or move a little faster on roller skates, in-line skates or bikes.

- Set up active play equipment in your backyard: a basketball hoop, a volleyball net, a jungle gym, or horseshoes. With a built-in sports center, you can catch 30 minutes of exercise time without leaving home.

- For cold weather days, there are both indoor and outdoor options. If you have room for the table, a brisk game of Ping-Pong can give you a real workout. Or bundle up and build a snowman.

- When all the relatives or friends get together, do more than sit and talk. Enjoy a softball game. Sponsor your own mini "Olympics" with races, broad jumps, bicycling, and other individual sports.

AN ACTIVE LIFESTYLE

44

ENJOY ACTIVE VACATIONS.

Whether your vacation is a weeklong trip or a weekend getaway, make it active and fun! Relaxing on a beach, snuggling into a deck chair on a cruise ship, or sitting by a stream doesn't provide much exercise. Balance your leisure time with plenty of exercise.

- Choose active leisure sports. If you hire a golf cart, a weekend golf outing probably won't offer much exercise. Neither will sitting with a fishing pole, cruising in a motor boat, or waiting in a hunting blind. Instead, walk the golf course, fish and cruise in a canoe or rowboat, and hike while you hunt.

- Enjoy an adventure vacation: whitewater rafting, backpacking, canoeing, or bicycling.

- Sign up for a family dude ranch or a tennis camp.

- Choose lodgings with recreational equipment and opportunities, such as a swimming pool, a boat house, hiking trails, and a well-equipped exercise room. Then use them!

OTHER FACTORS

45

KEEP YOUR HEALTHY WEIGHT.

Here's another issue that's dear to your heart—your healthy weight. Obesity is related to high blood cholesterol levels, heart disease, stroke, and high blood pressure, as well as some cancers, diabetes, gall bladder disease, and muscle and joint problems. Imagine carrying around a 25-pound sack of potatoes all the time; carrying excess body weight is even more of a strain on the body.

When weight comes down to a normal range, your risk of heart disease decreases. HDL cholesterol levels usually go up, and LDL cholesterol levels go down. Your heart is released from having to supply all that excess body fat with blood and can more efficiently supply the more vital organs—including itself.

What is your healthy weight? That's an individual matter. In part, it depends on where your body fat is located (see page 117), how much of your weight is fat, and whether you have a family history of heart disease. Carrying extra weight around the middle, may put you at greater risk than

OTHER FACTORS

extra weight on your hips or thighs. Check with your doctor if you are unsure about your healthy weight.

Diet + Exercise = Weight Control

Weight gain is in large part a matter of mathematics. By consuming more calories than you burn, your body stores the extra calories as fat, and you gain weight. If you consume fewer calories than you burn, your body draws from its stored supply, and you lose weight. Keep these hints in mind to tip the scales in your favor:

- *Set a reasonable goal.* A half to a pound per week is usually safe. A faster weight shift isn't healthy, especially if you must severely restrict the variety of foods or overall calories you consume.

- *Cut back on dietary fat.* Fat has 9 calories per gram, more than twice the amount that carbohydrate and protein have. Cutting back on fat may reduce your total calorie intake enough for weight loss. New research also suggests that calories from fat convert to body fat more easily than calories from carbohydrate and protein. (This should fit right in with your plan to lower cholesterol anyway.)

OTHER FACTORS

- *Set your calorie level for weight loss.* Use the chart in tip 7 as a guideline in figuring the number of calories you need. Women need at least 1,600 calories a day to get all the nutrients their bodies need. Unless you're under a doctor's supervision, don't eat less. With mostly lower-fat foods, that's the fewest servings recommended for each food group of the Food Guide Pyramid (see tip 6). With their height and larger body frame, most men consume more calories and still lose weight.

- *Consider your eating habits.* Do you always snack when you watch television whether or not you're hungry? Do you buy a doughnut during your morning break—just because you always do? Be aware of when and what you eat.

- *Add in exercise.* Controlling weight isn't just a matter of cutting calories and trimming fat in your food choices. Exercise is important to the equation, too. Physical activity burns calories. The more frequent and intense the exercise, the more calories you burn. Exercise increases the proportion of lean to fat body tissue, and lean muscle tissue burns more calories than fat even when it's at rest.

OTHER FACTORS

Where Is Your Fat?

To check your body shape, look in the mirror. To be more precise, measure your waist-to-hip ratio. First, measure your waist near your navel; relax and don't pull your stomach in! Second, measure your hips where they are the widest. Then, figure the waist-to-hip ratio by dividing your waist measurement by your hip measurement. If the total is above 1 (that is, your waist measurement is bigger than your hip measurement), your risk for heart disease and diabetes is greater.

Talk to your physician before you start exercising to lose weight if you have a lot to lose, if you're over age 45, if you've been sedentary, or if you have health problems. Even if none of these applies, checking with your doctor is probably a good idea; he or she can certainly give you some guidance.

Together, calorie control and physical activity help you achieve—and maintain—your healthy weight.

OTHER FACTORS

46

BE EXTRA CAREFUL AFTER MENOPAUSE.

Cardiovascular disease isn't just for men! After menopause, the risk of heart disease among women goes up dramatically! Statistics show that after age 50, a woman's risk of dying of cardiovascular disease is more than double her risk of dying of cancer. About 250,000 women die each year of heart disease.

What accounts for the dramatic change after menopause? Throughout a woman's life, the female reproductive hormone estrogen appears to keep HDL cholesterol levels high. However, at menopause, estrogen levels drop, and this protective effect declines, too. As a result, blood cholesterol levels often rise, and HDL cholesterol levels fall.

For many women, estrogen replacement therapy offers some benefit. The replacement hormones help lower LDL cholesterol levels in the same way the natural ones did before menopause. Whether women take estrogen or not, though, healthful food choices and exercise are as important as ever to the heart health of women over 50.

OTHER FACTORS

If you are approaching, experiencing, or past menopause, follow this health advice:

- Have your blood cholesterol and blood pressure checked at your regular physical examinations. If the numbers begin to climb, take an inventory of which behaviors you can change to keep the levels in check.

- Eat a healthful diet. Keep within your fat budget by choosing mostly nutritious, lower-fat foods.

- Make regular exercise part of your lifestyle—at least 30 minutes a day, most days of the week.

- If you smoke, quit. Smoking after menopause triples the risk of heart disease.

OTHER FACTORS

47

DRINK WINE IN MODERATION ONLY.

Will a daily glass of red wine protect you from heart disease? It probably won't hurt, but moderate drinking may offer no real benefit either. Some studies suggest a link between moderate drinking and heart benefits, but this evidence is not as strong as the evidence showing the benefits of quitting smoking, eating less fat, and engaging in moderate exercise. In fact, no professional health organizations currently recommend that anyone take up drinking to ward off heart disease.

Evidence links heavy drinking with heart disease, certain cancers, cirrhosis, stroke, and fetal damage (see tip 5). If you do drink alcoholic beverages, do so in moderation for your safety and good health. (Moderation is defined as one drink a day for women and two for men. A drink is 5 ounces of wine, 12 ounces of beer, or 1½ ounces of 80-proof liquor.)

OTHER FACTORS

48

SMOKERS—KICK THE HABIT!

Along with high blood cholesterol levels, high blood pressure, and inactivity, smoking is a major risk factor for heart disease that people can control. Smokers have lower HDL cholesterol levels than nonsmokers. By quitting, they can reverse the effect. So, if you are a smoker, kick the habit!

Those who quit need to watch what they eat. Ex-smokers often substitute food for cigarettes. The more they smoked, the more they tend to eat. To control the urge, drink plenty of water—8 to 10 cups daily to flush the body of nicotine. Don't linger over coffee or alcoholic drinks if these trigger the urge to smoke. If you nibble, try fruit, raw vegetables, pretzels, or other low-fat snacks. Better yet, take a walk or substitute some other form of exercise for a cigarette.

LIFETIME SUCCESS

49

ASK THE EXPERTS.

At least once a week, nutrition makes the news. With all you read and hear, how can you sift fact from fiction? How can you discern sound advice from preliminary research? How can you sort through seemingly conflicting recommendations? Just who's an expert?

Take a sensible approach. And don't believe all you read or hear.

- Be wary of attention-grabbing headlines. Read or listen well to learn the whole report.

- Find out the source. Does the expert represent a well-respected institution? Is the expert qualified to address the issue? Registered dietitians, nutrition scientists, and physicians are credible; entertainment stars are not.

- Avoid changing your diet too soon, especially after reports of a preliminary study. Wait until guidelines have been issued on the subject or at least until you can discuss it with your doctor or registered dietitian.

LIFETIME SUCCESS

- Remember the basics of good nutrition. Variety, moderation, and balance will always apply.

For information you can trust, contact your doctor or a registered dietitian. You can find a registered dietitian through your physician, local hospital, or clinic. The National Center for Nutrition and Dietetics of The American Dietetic Association (ADA) also has an inquiry hotline for consumers: 1-800-366-1655.

Other reliable nutrition resources in your community may include local food and nutrition extension agents, government agencies, and nutrition professionals at local offices of the American Heart Association, the American Diabetes Association, and some food commodity groups, such as the Dairy Council. The National Cholesterol Education Program is another fine source of information and materials.

To keep within your fat and cholesterol budget, you have plenty of help these days. Don't be afraid to ask for it.

LIFETIME SUCCESS

50

SAVOR THE RESULTS.

With 50 tips under your belt, you're ready to start on your lighter lifestyle. Start implementing these strategies slowly. Gradually incorporating healthy habits, rather than making radical changes all at once, will help you stick with it. For example, start making one vegetarian dinner a week and exercising a few times a week; implement a few more ideas that you learned in this book when you feel comfortable with this level. Soon you'll realize that lowering your fat and cholesterol isn't a chore at all. In fact, you may even find that low-fat eating, regular exercise, and wise lifestyle choices will make you feel great about yourself and your health.

After a while, go back and rate your plate again (tip 4). Test your exercise quotient once more, too (tip 36). What changes have you made since you first scored yourself? Maybe you're learning to use low-fat cooking techniques or taking the stairs instead of the elevator. Pat yourself on the back for all of the **N**s that are now **S**s and all the **S**s that are now **A**s.

LIFETIME SUCCESS

Stay in touch with your numbers. Remember to have your blood cholesterol levels checked regularly. If the number starts to get out of your healthy range, go back through these tips and see what other healthy habits you can adopt to help you get the levels back in line. Also, remember to keep your calorie target, fat budget, and healthy weight in mind so you can stay on track.

Healthful eating and exercise aren't just short-term strategies. Make a long-term commitment to the lifestyle changes that will keep you feeling in top form. Remember that healthy eating averages out over several days. So don't feel guilty if you miss your target one day; just compensate over the next few days. You're not making these changes to deprive yourself; you're free to enjoy the foods you always have. Moderation is not deprivation, and exercise need not be a chore; both are ways to maximize your potential.

INDEX

alcohol, 25, 120
arteries, 10–11, 95

baking, 60, 67, 74–75
blood pressure
 calcium and, 53
 exercise and, 103
 obesity and, 114
 smoking and, 121
 sodium and, 24
 in women, 118–119

calcium, 53–55, 64, 85
 on food labels, 39
calories, 23, 26–27
 burning of, 94, 97, 103–104, 107–108, 115–116
 conversion of, 40
 from fat, 12, 35, 55
 requirements, 27, 28, 32, 115–116
cancer
 high-fat diet and, 9
 fiber and, 78, 79
 obesity and, 114
 risk of, 118
 vegetables and, 48
carbohydrates
 complex, 23, 48, 80, 91
 on food labels, 39, 40
 sources of, 48, 51, 91
cardiovascular disease, 17, 24, 118
cholesterol
 blood, 7, 9–12, 13, 17–18, 79, 114, 118, 119

cholesterol *(continued)*
 budget, 30–33, 123
 dietary, 7, 11, 13, 21, 22, 30–33, 74
 exercise and, 95
 fiber and, 78–79
 on food labels, 36–40, 41, 56–57
 meat and, 43, 45, 72
 reduction, 7, 22
 screening, 17–18, 119, 125
 smoking and, 121
 sources, 13

dairy products, 9, 13, 35, 53–55
 low-fat, 54–55
diabetes
 body fat and, 22, 114, 117
diet, 9, 11, 19–29, 30, 122
 fat in, 60
 high-fat, 9–11
 low-fat, 22
 vegetarian, 80–81
 weight control and, 115–116
dietitian, 80–81, 122, 123

estrogen, 118
exercise, 94–113, 115–116, 124, 125
 aerobic, 95, 109–111
 anaerobic, 95, 109–110

INDEX

exercise *(continued)*
  appetite and, 96
  blood cholesterol and, 7, 11, 95
  stretching, 98, 110–111
  weight-bearing, 97
  weight control and, 22, 94–95, 96–97, 115–116

fat, 14–15
  blood cholesterol and, 9–12
  body, 14, 94, 97, 103–104, 109, 114–117
  budget, 30–33, 34–35, 72, 119, 123, 125
  calories from, 12, 35, 55
  dietary, 7–8, 14, 22, 26–27
  disease and, 9, 114, 117
  on food labels, 37–40, 41–42
  monounsaturated, 11–12
  polyunsaturated, 11–12, 41–42
  saturated, 11–12, 27, 30, 31–32, 42, 56–57
  sources of, 9, 41–47, 49, 52, 53–55, 75, 80, 83, 84
  triglycerides, 15, 17, 18, 46
  vitamins and, 7–8, 14

fatty acids, 15–16
  essentials, 15–16, 54
  monounsaturated, 15, 90
  omega-3, 16, 46
  omega-6, 16
  polyunsaturated, 15
  saturated, 15
  trans, 12, 16, 37, 41

fiber, 23, 39, 48, 52, 78–79, 80, 84,

fish, 26, 29, 46–47
  portions of, 72
  preparation of, 46–47, 61–62, 69–70, 91

Food Guide Pyramid, 26–29, 38, 56, 83, 86, 116

heart attack, causes, 10–11

heart disease
  body fat and, 22, 114, 117
  causes, 10–11, 107
  risk for, 9, 17–18, 22, 24, 46, 48, 114, 117, 118, 120, 121

heart, exercise and, 95–96, 103, 104, 109–110, 111, 118–119

heart rate, 101–102, 109–111

high blood pressure, 24, 53, 114
  smoking and, 121

INDEX

hydrogenation, 12, 16, 41, 42

labels. *See* **Nutrition Facts panel.**
legumes, 29, 79, 80–81, 93
lipoproteins, 10–11, 13–14, 16
  high-density, 10–11, 13–14, 17–18, 95, 114, 118, 121
  low-density, 10–11, 14, 17–18, 95, 114, 118

meat, 26, 29, 43–45
  labeling of, 36, 37, 43–44
  portions of, 29, 72–73, 83, 86
  preparation of, 45, 58, 59, 61–62, 65, 68, 69–70
menopause, 118–119
minerals, 23, 39

Nutrition Facts panel, 36–40, 52, 57

obesity, 9, 22, 114
oils, 9–10, 15–16, 41–42
  hydrogenated, 12, 16, 41, 42
  tropical, 9–10, 41–42, 52

osteoporosis, 53, 97

poultry, 26, 29, 43–45
  labeling, 36, 37, 43–44
  portions of, 29
  preparation of, 43–45, 59, 61–62, 65, 68, 69–70
protein, 80, 115
  complete, 80, 93
  on food labels, 39
  sources of, 43–45, 74

serving size, 29, 38, 72–73, 86
smoking, 11, 119, 121
sodium, 24–25, 39, 82
stress, 96, 98, 103
stroke
  causes, 11
  risk for, 114

vegetables, 23, 26, 27, 29, 49–50, 80, 84
  preparation of, 49–50, 60–61
  sauces from, 64
vitamins, 7–8, 14, 23, 39, 48, 60
  A, 7, 14, 23, 48
  C, 23, 48, 49
  D, 7, 14
  E, 7, 14
  K, 7, 14

weight control, 22, 94–95, 97, 114–117, 125